DATE DUE		

THE FACTS ON FILE
ALGEBRA
HANDBOOK

DEBORAH TODD

Facts On File, Inc.

Facts On File
132 West 31st Street
New York NY 10001

Library of Congress Cataloging-in-Publication Data

Todd, Deborah.
     The facts on file algebra handbook/Deborah Todd.
         p. cm. — (The facts on file science handbooks)
Includes bibliographical references and index.
     ISBN 0-8160-4703-0
     1. Algebra—Handbooks, manuals, etc. I. Title. II. Facts on File science
library.
     QA159.T63 2003
     512—dc21   2002154644

Facts On File books are available at special discounts when purchased in bulk quantities for businesses, associations, institutions, or sales promotions. Please call our Special Sales Department in New York at 212/967-8800 or 800/322-8755.

You can find Facts On File on the World Wide Web at
http://www.factsonfile.com

Cover design by Cathy Rincon
Illustrations by Anja Tchepets and Kerstin Porges

Printed in the United States of America

VB Hermitage 10 9 8 7 6 5 4 3 2 1

This book is printed on acid-free paper.

For Jason,
the light of my life

For Rob, Jennifer, Drena,
Mom, and Dad
for everything you are to me

For Jeb,
more than you'll ever know

CONTENTS

ACKNOWLEDGMENTS

This is the part of the book where the author always writes "this book would not have been possible without the help of the following people…" and it's true. In this case, many generous people have touched the making of this book in one way or another. My sincere gratitude and deep appreciation is offered to the following wonderful souls for their contributions in helping make this book a reality: Matt Beucler, for the road map, and because everybody needs a coach and you have been the best; and John Chen, for letting me figure it out by myself those many years ago in Hawaii, and for introducing me to Matt. Sarah Poindexter, for stating very simply what was real, this book became real because of you. Roger and Elizabeth Eggleston, who have contributed more time than anyone should ever be asked to, and more support than anyone could possibly imagine. David Dodd, reference librarian extraordinaire at the Marin County Civic Center Public Library. Heather Lindsay, of the Emilo Segrè Visual Archives of the American Institute of Physics, for the incredible help with photos, you saved me. Chris Van Buren and Bill Gladstone, of Waterside Productions, and the amazing Margot Maley Hutchison for stepping into the fray and agenting with such finesse and spirit. Frank Darmstadt, of Facts On File, a saint of an editor and the absolutely most patient man I have ever encountered in my life. You are one of a kind, I am certain of it. The support network of the famous Silicon Valley breakfast club, WiWoWo, especially Sally Richards, Carla Rayachich, Donna Compton, Renee Rosenfeld, Lucie Newcomb, Silva Paull (also of Gracenet fame), Liz Simpson, Joyce Cutler, et al., you have been with me through the entire time of this adventure, and yes, it's finally finished! Madeline DiMaggio, the world's best writing coach and a dear friend; Kathie Fong Yoneda, a great mentor and friend in all kinds of weather; Pamela Wallace, writer and friend extraordinaire who has gone above and beyond for me in all ways; and for the three of you for introducing me to the three of you. Gregg Kellogg for his incredible selfless research, late-night readings, and the dialogues about mathematicians. John Newby, who helped me keep moving forward. Rob Swigart, for the support, in so many ways, in research, time, and things too many to mention here, without whom I could not have done this on a number of levels, including the finite and infinite pieces of wisdom you have shared with me. A very special thank you to my dear friend and soul-sister Jennifer Omholt for keeping me laughing through the most bizarre circumstances that could ever happen to anyone while writing a book. My most heartfelt thank you and love to Jeb Brady, whose complete love and support, and total belief in me, gives me the absolute freedom to write and live with passion, more than you'll ever know. And my deepest thank you and love to my son, Jason Todd, whose genuine encouragement, understanding and acceptance of a writer's life, and sincere happiness for me in even my smallest accomplishments, is exceeded only by his great soul and capacity for love.

INTRODUCTION

The mathematics that we teach and learn today includes concepts and ideas that once were pondered only by the most brilliant men and women of ancient, and not so ancient, times. Numbers such as 1,000, for example, or two, or zero, were at one time considered very abstract ideas. There was a time when a quantity more than two or three was simply called "many." Yet we have grown up learning all about quantities and how to manipulate them. We teach even young children the concept of fractions as we ask them to share, or divide, their candy between them. Today, in many ways, what used to be stimulating thought for only the privileged few is now considered child's play.

Yet scholars, philosophers, scientists, and writers of the past have spent lifetimes devising ways to explain these concepts to benefit merchants, kings, and countries. The idea that two items of different weight could fall to the Earth at the same rate was, in its time, controversial. Creating calculations that pointed to the fact that the Earth revolved around the Sun was heresy. Mathematicians have, in fact, been beheaded by kings, imprisoned by churches, and murdered by angry mobs for their knowledge. Times have changed, thankfully. It is fair to say we have come a long way.

This book is designed to help you come even further in your understanding of algebra. To start with, there is a lot of algebra that you already know. The Additive Identity Property, the Commutative Property of Multiplication, the Multiplicative Property of Equality, and the Zero Product Theorem are already concepts that, while you might not know them by name, are in your personal database of mathematical knowledge. This book will help you identify, and make a connection with, the algebra that you already do know, and it will give you the opportunity to discover new ideas and concepts that you are about to learn.

This book is designed to give you a good broad base of understanding of the basics of algebra. Since algebra plays such an integral role in the understanding of other parts of mathematics, for example, algebraic geometry, there is naturally some crossover of terms. As you become interested in other fields of mathematics, whether on your own or through formal study, you have the resources of *The Facts On File Geometry Handbook* and *The Facts On File Calculus Handbook* for your referral.

The foundation of this book is the belief that everyone deserves to have algebra made easy and accessible to them. The *Facts On File Algebra Handbook* delivers on this idea in an easy-to-access resource, providing you with a glossary of terms, an expanded section of charts and tables, a chronology of events through time, and a biography section on many of the people who have dedicated at least a portion of their lives to enrich ours with a better understanding of mathematics. In the spirit of their dedication, this book is dedicated to you.

GLOSSARY

This section is your quick reference point for looking up and understanding more than 350 terms you are likely to encounter as you learn or rediscover algebra. What is a radicand, a quotient, a polygon? What is the difference between median and mean? What are a monomial, a binomial, and a polynomial? Many glossary entries are elaborated on in the Charts and Tables section of this book, where you will find a more in-depth explanation of some of the terms and their calculations.

BIOGRAPHIES

The biography section is full of colorful characters. Charles Babbage hated street musicians. Girolamo Cardan slashed a man's face because he thought he was being cheated at cards. Evariste Galois was simultaneously the most brilliant and the most foolish man in the history of mathematics. There are also many people listed here who offer great inspiration. The brilliant Sir Isaac Newton did not start school until he was 10 years old, and he was 20 before he ever saw his first book on mathematics. Andrew Wiles decided at the age of 10 that he was going to solve Fermat's Last Theorem, and he did! Florence Nightingale calculated that if hospital conditions did not improve, the entire British army would be wiped out by disease. Her calculations changed the nature of medical care.

There are more than 100 brief biographies that give you a glimpse at the people who have made important contributions to the art and science of mathematics, especially in algebra. Use this as a starting point to find out more about those who particularly interest you. The Recommended Reading section in the back of the book will help guide you to other great resources to expand your knowledge.

A word about dates: Throughout time, calendaring and chronicling dates has been inconsistent at best. A number of different dates are recorded in research on a variety of people, for a variety of reasons, and some of the dates you find in this book will not match with some you might find in other references. In many cases, no documentation exists that gives a precise date for someone's birth. Often, dates have been calculated by historians, and many historians disagree with each other's calculations. In addition, many countries have used different calendar systems, making it impossible to have a date that anyone agrees on for any given event. For example, December 24 in one calendar system might be calculated to be January 7 in another. The dates used in this section reflect the most common aggregate of dates considered to be accurate for any individual listed here.

CHRONOLOGY

Did you know that the famous Egyptian Rosetta Stone helped play a part in our understanding of ancient mathematics? Or that Galileo Galilei died the same day Sir Isaac Newton was born? Our history of algebra dates from ancient times, through the Renaissance, to the present day, spanning nearly 4,000 years of events. These remarkable contributions of the past have made it possible to develop everything from the chaos theory to telephones and computers.

CHARTS AND TABLES

The Glossary is the best place to get a quick answer on the definition of a word or phrase. The Charts and Tables section is your best resource for some in-depth examples. You will also find items here that are organized in a precise way for a quick reference on specific information you might need, such as the different types of numbers, the kinds of plane figures, the characteristics of different triangles, and how to calculate using F.O.I.L. or P.E.M.D.A.S. There is an extensive section on measurements and their equivalents, another on theorems and formulas, and still another on mathematical symbols that will be helpful as you delve into your study of algebra.

RECOMMENDED READING

This section offers some suggestions on where to get more information on the topics found in this book. They run the gamut from historical perspectives, such as *A Short Account of the History of Mathematics,* by W. W. Rouse Ball, to textbooks like *Forgotten Algebra,* by Barbara Lee Bleau. Website resources, which have the ability to change in an instant, are also listed as reference.

SECTION ONE
GLOSSARY

abscissa On an *(x, y)* GRAPH, the *x* coordinate is the *abscissa,* and the *y* coordinate is the ORDINATE. Together, the abscissa and the ordinate make the coordinates.

absolute value (numerical value) The number that remains when the plus sign or minus sign of a SIGNED NUMBER is removed. It is the number without the sign. The symbol for absolute value is indicated with two bars, like this: | |.
　　See also SECTION IV CHARTS AND TABLES.

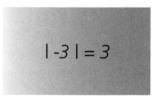

Absolute value

abundant number Any number whose FACTORS (excluding the number itself), when added up, equal more than the number itself. For example, the factors for the number 12 are 1, 2, 3, 4, and 6. When these numbers are added, the SUM is 16, making 12 an abundant number.

acute angle Any ANGLE that measures less than 90°.

acute triangle A TRIANGLE in which all angles are less than 90°.
　　See also SECTION IV CHARTS AND TABLES.

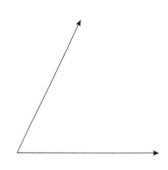

Acute angle

addend Any number that is added, or is intended to be added, to any other number or SET of numbers.

Additive Identity Property Any number added to ZERO equals the number itself. For example, $3 + 0 = 3$.
　　See also SECTION IV CHARTS AND TABLES.

additive inverse A number that is the opposite, or inverse, or negative, of another number. When expressed as a VARIABLE, it is written as $-a$, and is read as "the opposite of *a*," "the additive inverse of *a*," or "the negative of *a*."

Additive Inverse Property For every REAL NUMBER *a*, there is a real number $-a$ that when added to *a* equals ZERO, written as $a + (-a) = 0$.
　　See also SECTION IV CHARTS AND TABLES.

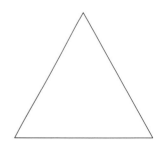

Acute triangle

Additive Property of Equality If two numbers are equal, for example if $a = b$, when they are both added to another number, for example *c*, their SUMS will be equal. In EQUATION form, it looks like this: if $a = b$, then $a + c = b + c$.
　　See also SECTION IV CHARTS AND TABLES.

Additive Property of Inequality If two numbers are not equal in value, for example if $a < b$, and they are added to another number, for example

Adjacent angle

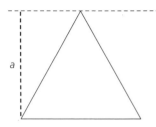

Altitude

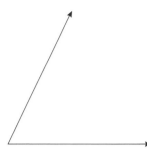

Angle

c, their SUMS will not be equal. The EXPRESSION looks like this: if $a < b$, then $a + c < b + c$.

  See also SECTION IV CHARTS AND TABLES.

adjacent angle (contiguous angle) Either of two ANGLES that share a common side and VERTEX.

altitude In a figure such as a TRIANGLE, this is the DISTANCE from the top of the PERPENDICULAR line to the bottom where it joins the BASE, and is usually indicated with the letter *a.* In a SOLID figure, such as a PYRAMID, this is the perpendicular distance from the VERTEX to the base.

angle The shape formed by two lines that start at a common point, called the VERTEX.

angle of depression The ANGLE formed when a HORIZONTAL LINE (the PLANE) is joined with a descending line. The angle of depression is equal in VALUE to the ANGLE OF ELEVATION.

angle of elevation The ANGLE formed when a HORIZONTAL LINE (the PLANE) is joined with an ascending line. The angle of elevation is equal in VALUE to the ANGLE OF DEPRESSION.

antecedent The first TERM of the two terms in a RATIO. For example, in the ratio 3:5, the first term, 3, is the antecedent.

  See also CONSEQUENT.

apothem The PERPENDICULAR DISTANCE of a LINE SEGMENT that extends from the center of a REGULAR POLYGON to any side of the POLYGON.

arc A segment of a curved line. For example, part of a CIRCUMFERENCE.

area The amount of surface space that is found within the lines of a two-dimensional figure. For example, the surface space inside the lines of a TRIANGLE, a CIRCLE, or a SQUARE is the area. Area is measured in square units.

  See also SECTION IV CHARTS AND TABLES.

arithmetic sequence (linear sequence) Any SEQUENCE with a DOMAIN in the SET of NATURAL NUMBERS that has a CONSTANT DIFFERENCE between the numbers. This difference, when graphed, creates the SLOPE of the numbers. For example, 1, 3, 5, 7, 9, __ is an arithmetic sequence and the common difference is 2.

  See also SECTION IV CHARTS AND TABLES.

arithmetic series A SERIES in which the SUM is an ARITHMETIC SEQUENCE.

Associative Property of Addition When three numbers are added together, grouping the first two numbers in parentheses or grouping the last two numbers in parentheses will still result in the same SUM. For example, $(1 + 3) + 4 = 8$, and $1 + (3 + 4) = 8$, so $(1 + 3) + 4 = 1 + (3 + 4)$.
See also SECTION IV CHARTS AND TABLES.

Associative Property of Multiplication When three numbers are multiplied together, grouping the first two numbers in parentheses or grouping the last two numbers in parentheses will still result in the same PRODUCT. For example, $(2 \cdot 3) \cdot 4 = 24$, and $2 \cdot (3 \cdot 4) = 24$, so $(2 \cdot 3) \cdot 4 = 2 \cdot (3 \cdot 4)$.
See also SECTION IV CHARTS AND TABLES.

average (mean) See MEAN.

axes More than one AXIS.

axiom (postulate) A statement that is assumed to be true without PROOF.

Axiom of Comparison For any two quantities or numbers, for example, a and b, one and only one condition can be true; either $a < b$ or $a = b$ or $b < a$.

axis An imaginary straight line that runs through the center of an object, for example a CIRCLE or a cylinder.

axis of symmetry of a parabola The VERTICAL LINE, or AXIS, which runs through the VERTEX of a PARABOLA, around which the points of the curve of the parabola on either side of the axis are symmetrical.

bar graph (bar chart) A chart that uses RECTANGLES, or bars, to show how the quantities on the chart are different from each other.

base
1. In an EXPRESSION with an EXPONENT, the base is the number that is multiplied by the exponent. For example, in the expression 5^2, 5^3, 5^n, the base is 5.
2. In referring to a number system, the base is the RADIX.
3. In referring to a figure, such as a TRIANGLE, the base is the side on which the figure sits, and is usually indicated with the letter b.

bel A unit of measure of sound, named after Alexander Graham Bell, that is equal to 10 DECIBELS.

bi- Two.

binary A number system that uses only 0 and 1 as its numbers. In a binary system, 1 is one, 10 is two, 11 is three, 100 is four, and so on. A

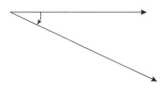

Angle of depression

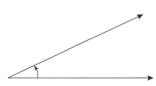

Angle of elevation

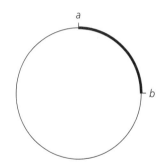

Arc

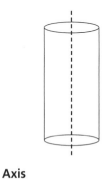

Axis

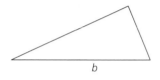

Base

binary system is a base 2 system that is typically used in computers and in BOOLEAN ALGEBRA.

binomial A POLYNOMIAL that has just two TERMS, in other words, is an EXPRESSION that consists of a string of just two MONOMIALS. For example, $\frac{1}{2}x + 5xy^2$.

 See also SECTION IV CHARTS AND TABLES.

Boolean algebra Named after GEORGE BOOLE, this type of computation is based on logic and logical statements, and is used for SETs and diagrams, in PROBABILITY, and extensively in designing computers and computer applications. Typically, letters such as $p, q, r,$ and s are used to represent statements, which may be true, false, or conditional.

 See also SECTION IV CHARTS AND TABLES.

branches of a hyperbola The two curves of a HYPERBOLA found in two separate QUADRANTS of the GRAPH.

canceling Dividing the NUMERATOR and DENOMINATOR of a FRACTION by a COMMON FACTOR, usually the HIGHEST COMMON FACTOR.

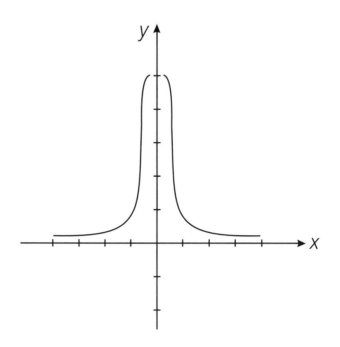

Branches of a hyperbola

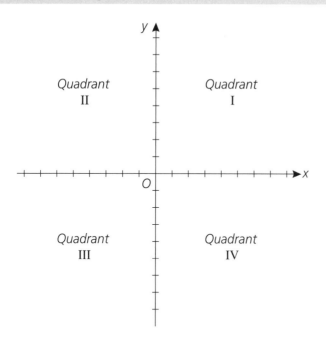

Cartesian coordinate system A GRAPH consisting of a two-dimensional PLANE divided into quarters by the PERPENDICULAR AXES, the x-axis and y-axis, for the purpose of charting COORDINATES.

chord Any LINE SEGMENT that joins two points of a CIRCLE without passing through the center.

circle A closed PLANE figure that is made of a curved line that is at all points EQUIDISTANT from the center.

circulating decimal *See* REPEATING DECIMAL.

circumference The DISTANCE around the curved line of a CIRCLE. The formula for calculating the circumference of a circle is $C = 2\pi r$.

    See also SECTION IV CHARTS AND TABLES.

coefficient The quantity in a TERM other than an EXPONENT or a VARIABLE. For example, in the following terms, the variables are x and y, and the numbers 3, 5, 7, and 9 are the coefficients of each term: $3x$, $5y$, $7xy$, $9\pi x^2$.

coefficient matrix A MATRIX (rectangular system of rows and columns) used to show the VALUES of the COEFFICIENTS of multiple EQUATIONS. Each row shows the SOLUTIONS for each equation. If there are four

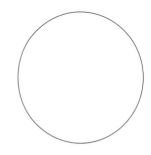

Chord

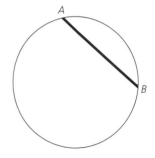

Circle

equations, and each equation has three solutions, the matrix is called a 4×3 matrix (four by three matrix).

collinear Two or more points that are located on the same line.

combining like terms Grouping LIKE TERMS together to SIMPLIFY the calculation of an EXPRESSION OR EQUATION. For example, in $3x + 7x - y$, the like terms of $3x$ and $7x$ can be combined to simplify, creating the new expression $10x - y$.

common denominator The same value or INTEGER in the DENOMINATOR of two or more FRACTIONS. For example, in $\frac{3}{4} + \frac{2}{4}$ the common denominator is 4. In $\frac{a}{b} + \frac{d}{b}$, the common denominator is b. In $\frac{2}{6} + \frac{3}{6} + \frac{4}{6}$, the common denominator is 6. Fractions can be added and subtracted when they have a common denominator.

See also SECTION IV CHARTS AND TABLES.

common divisor *See* COMMON FACTOR.

common factor (common divisor, common measure) Any number that can be divided into two other numbers without leaving a REMAINDER. For example, a common factor of the numbers 6 and 12 is 3. Another common factor is 2.

See also GREATEST COMMON FACTOR.

common fraction (simple fraction) Any FRACTION that has a WHOLE NUMBER as the NUMERATOR, and a whole number as the DENOMINATOR.

See also SECTION IV CHARTS AND TABLES.

common measure *See* COMMON FACTOR.

common multiple Any number that is a MULTIPLE of two or more numbers. For example, 4, 8, 12, and 16 are common multiples of both 2 and 4, and the numbers 12, 24, and 36 are common multiples of 3 and 4.

Commutative Property of Addition Two numbers can be added together in any order and still have the same SUM. For example, $1 + 3 = 4$, and $3 + 1 = 4$.

See also SECTION IV CHARTS AND TABLES.

Commutative Property of Multiplication Two numbers can be multiplied together in any order and still have the same PRODUCT. For example, $2 \times 3 = 6$, and $3 \times 2 = 6$.

See also SECTION IV CHARTS AND TABLES.

complementary angles Two ANGLES that, when summed, equal $90°$.

completing the square Changing a QUADRATIC EQUATION from one form to another to solve the EQUATION. The standard form is $y = ax^2 + bx + c$, and the VERTEX form is $y - k = a(x - h)^2$. To complete the square on a POLYNOMIAL of the form $x^2 + bx$, where the COEFFICIENT of x^2 is 1, the THEOREM is to add $(\frac{1}{2}b)^2$.

complex fraction (compound fraction) Any FRACTION that has a fraction in the NUMERATOR and/or the DENOMINATOR.

    See also SECTION IV CHARTS AND TABLES.

complex number The resulting number from the EXPRESSION $a + bi$. The VARIABLES a and b represent REAL NUMBERS, and the variable i is the IMAGINARY NUMBER that is the SQUARE ROOT of -1.

composite number Any integer that can be divided *exactly* by any POSITIVE NUMBER other than itself or 1. For example, the number 12 can be divided exactly by 4, 3, 2, or 6. Other composite numbers include 4, 6, 8, 9, 10, 12, 14, 15, 16, 18, 20, and so on.

    See also SECTION IV CHARTS AND TABLES.

compound fraction *See* COMPLEX FRACTION.

compound number Any quantity expressed in different units. For example, 6 feet 2 inches, 8 pounds 1 ounce, 5 hours 15 minutes, and so on.

compound quantity Any quantity consisting of two or more TERMS connected by a + or – sign. For example, $3a + 4b - y$, or $a - bc$.

compound statement (compound sentence) Two sentences combined with one of the following words: *or, and.* In combining SETs, the word *or* indicates a UNION between the sets; the word *and* indicates an INTERSECTION between the sets.

concave A rounded surface that curves inward.

conditional equation Any EQUATION in which the VARIABLE has only certain specific VALUEs that will make the equation true.

conditional statement Any statement that requires one matter to be true for the subsequent matter to be true. Also called an "If, then" statement, and often used in BOOLEAN ALGEBRA as "if p, then q," written as $p \rightarrow q$.

conjecture To hypothesize about a conclusion without enough evidence to prove it.

consecutive integers Counting by one, resulting in INTEGERs that are exactly one number larger than the number immediately preceding.

Complex fraction

Concave

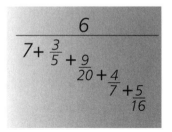

Continued fraction

Convex

consequent The second TERM of the two terms in a RATIO. For example, in the ratio 3:5, the second term, 5, is the consequent.
See also ANTECEDENT.

constant A value that does not change and is not a VARIABLE.

contiguous angle See ADJACENT ANGLE.

continued fraction Any FRACTION with a NUMERATOR that is a WHOLE NUMBER—and a DENOMINATOR that is a whole number plus a fraction, which fraction has a numerator that is a whole number and denominator that is a whole number plus a fraction, and so on.

continuous graph Any GRAPH in which the entire line of the graph is one consecutive, or continuous, line.

converse The inversion of a proposition or statement that is assumed true, based on the assumed truth of the original statement. For example, if $A = B$, the CONVERSE is $B = A$. In BOOLEAN ALGEBRA, for the CONDITIONAL STATEMENT "if p, then q," written as $p \rightarrow q$, the converse is "if q, then p," written as $q \rightarrow p$.

convex A spherical surface that curves outward.

coordinates The numbers in an ORDERED PAIR. The x-coordinate is always the first number, and the y-coordinate is always the second number. For example, in the coordinates $(5, -2)$, 5 is the x-coordinate, and -2 is the y-coordinate.
See also ABSCISSA and ORDINATE.

cross multiplication Multiplying the NUMERATOR of one FRACTION by the DENOMINATOR of another fraction.

cube 1. The third POWER of any number.
2. A SOLID three-dimensional shape with six sides, each side having the exact same measurements as the others.
See also SECTION IV CHARTS AND TABLES.

cubed A BASE number that is raised to the third POWER.

cubic Third-degree term. For example, x^3.

cubic equation An EQUATION that contains a TERM of the third degree as its highest POWERed term. An equation in which the highest power is an x^2 is a SECOND-DEGREE EQUATION, an x^3 is a THIRD-DEGREE EQUATION, an x^4 is a fourth-degree equation, x^5 is a fifth-degree equation, and so on.

cubic unit The measurement used for the VOLUME of a SOLID. For example, cubic meter, cubic yard, cubic inch.

decagon A 10-sided POLYGON.

decibel A UNIT that expresses the intensity of sound as a FRACTION of the intensity of a BEL. One decibel is equal to $\frac{1}{10}$ of a bel. The symbol for decibel is dB.

decimal The decimal system is a number system based on 10s. Usually, decimals refer to *decimal fractions,* so $3\frac{1}{10}$ is written as 3.1, $76\frac{95}{100}$ is written as 76.95, and so on.

decimal fraction Any FRACTION with a DENOMINATOR that is a power of 10, such as $\frac{1}{10}, \frac{37}{100}, \frac{629}{1000}$, and so on. This kind of fraction is usually written in DECIMAL form, for example, $\frac{1}{10}$ is written as .1, $\frac{37}{100}$ is written as .37, and $\frac{629}{1000}$ is written as .629.

decimal point A dot used in base 10 number systems to show both INTEGER and FRACTION values. The numbers to the left of the dot are the integers, and the numbers to the right of the dot are the fractions. For example, 0.4, 3.6, 1.85, 97.029, and so on.

deficient number (defective number) Any number whose FACTORS (excluding the number itself), when added up, equal less than the number itself. For the number 14, the factors are 1, 2, and 7. When these numbers are added, the SUM is 10, making 14 a deficient number.

degree of a polynomial The degree of the highest EXPONENT in a POLYNOMIAL. For example, in the polynomial $\frac{1}{2}x + 5xy^2 + \pi$, the degree is 2 because the highest exponent of 2 is found in the TERM $5xy^2$.
    See also SECTION IV CHARTS AND TABLES.

denominator The number in a FRACTION that is below the division line, showing how many equal parts the WHOLE NUMBER has been divided into. For example, in $\frac{1}{2}$, the denominator is 2, meaning that the whole has been divided into 2 equal parts. In $\frac{3}{4}$ the denominator is 4, and the whole has been divided into four equal parts. In $\frac{7}{8}$ the denominator is 8, in $\frac{9}{16}$ the denominator is 16, and in $\frac{97}{100}$ the denominator is 100. ZERO is never used as a denominator.
    See also SECTION IV CHARTS AND TABLES.

dependent variable The VARIABLE that relies on another variable for its VALUE. For example, in $A = \pi r^2$, the value of the AREA depends on the value of the RADIUS, so A is the dependent variable.

description method of specification (rule method) The method in which the elements, or MEMBERS, of a SET are described. For example, $A = \{$the EVEN NUMBERS between 0 and 10$\}$. An

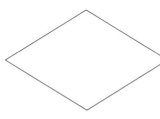

Diameter

Diamond

alternative method of listing elements of a set is the list, or ROSTER METHOD of SPECIFICATION.

diagonal Any straight LINE SEGMENT that joins two nonadjacent or nonconsecutive vertices on a POLYGON, or two vertices on different faces on a POLYHEDRON.

diameter The length of a LINE SEGMENT that dissects the center of a CIRCLE, with the ends of the line segment at opposite points on the circle.

diamond A QUADRILATERAL that has two OBTUSE ANGLES and two ACUTE ANGLES.
  See also SECTION IV CHARTS AND TABLES.

difference The total obtained by subtracting one number or quantity from another. For example, in $103 - 58 = 45$, the total 45 is the difference.

difference of two cubes A formula used to FACTOR two CUBED BINOMIALS into two POLYNOMIALs of the SUM of the CUBE roots times the DIFFERENCE of the cube roots, written as $x^3 - y^3 = (x - y)(x^3 + xy + y^3)$.

difference of two squares A formula used to FACTOR two SQUARED BINOMIALs into the SUM of the SQUARE ROOTs times the DIFFERENCE of the square roots, written as $x^2 - y^2 = (x + y)(x - y)$.

dimension The measurement of an object or figure, including length, width, depth, height, mass, and/or time.

directly proportional The change in the value of a VARIABLE as it relates to the change in the value of another variable in a DIRECT VARIATION FORMULA. For example, in $A = \pi r^2$, the value of the AREA changes as a direct result of changes in the value of the RADIUS, so A is directly proportional to r^2.

direct variation formula Any formula in which the value for one VARIABLE is dependent on the value of another variable. For example, the AREA of a CIRCLE, $A = \pi r^2$. This formula is expressed as $y = kx^n$.

discontinuous graph Any GRAPH in which the lines of the graph are not one consecutive, or continuous, line.

discrete graph Any GRAPH in which the points are not connected.

Discriminate Theorem A THEOREM used to determine the number of ROOTS of a QUADRATIC EQUATION.
  See also SECTION IV CHARTS AND TABLES.

distance 1. Measurement that is equal to rate of speed multiplied by time.
  2. The length between two or more points.

Discontinuous graph

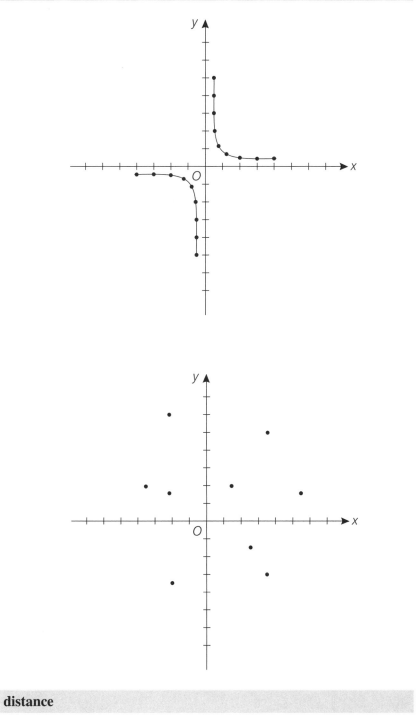

Discrete graph

Distributive Property For any REAL NUMBERS *a, b,* and *c,* the SUM of two numbers times a third is equal to the sum of each number times the third; written as $a(b + c) = ab + ac$ or $(a + b)c = ac + bc$.
See also SECTION IV CHARTS AND TABLES.

dividend Any number that is to be divided by another number. For example, in $4 \div 2$, 4 is the dividend, in $15 \div 3$, 15 is the dividend, in $\frac{1}{2}$, 2 is the dividend, in $\frac{7}{8}$, 8 is the dividend, and so on. The number that divides the dividend is the DIVISOR.

divisor The number that divides the DIVIDEND. For example, in $4 \div 2$, 2 is the divisor.
See also NUMERATOR.

dodecagon A 12-sided POLYGON.

dodecahedron A 12-sided POLYHEDRON in which all 12 faces are PENTAGONS.
See also SECTION IV CHARTS AND TABLES.

domain *See* REPLACEMENT SET.

double inequality A MATHEMATICAL SENTENCE that contains exactly two identical INEQUALITY symbols. For example, $3 \geq x \geq 12$; or $-5 < y < 7$.

elements *See* MEMBERS.

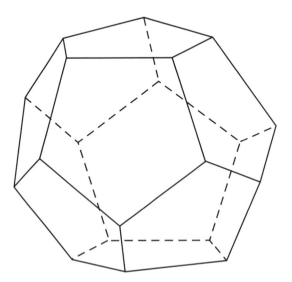

Dodecahedron

ellipses A series of three dots, . . . , used to indicate that a series of numbers continues on in the same pattern. For example, {2, 4, 6, 8, . . .} indicates that the SET continues on with EVEN NUMBERS; {10, 20, 30, 40, . . .} indicates that the set continues on by tens.

empirical probability The PROBABILITY of a future event happening, given the actual data of the event happening in the past.

empty set (null set) A SET that contains no elements. For example, if $A =$ {the ODD NUMBERs in the group 2, 4}, then A has no elements in it, and is considered an empty set. An empty set is designated with the symbol $\varnothing$. Since an empty set has no elements in it, the empty set is considered a SUBSET of every set.

Empty set

equality (equals) Exactly the same VALUE in quantity between two EXPRESSIONS, usually shown with the symbol =.

equation Any MATHEMATICAL SENTENCE that contains an "equals" (=) symbol. For example, $3n$ is an EXPRESSION, $3 + n < 12$ is a mathematical sentence, and $3 + n = 12$ is an equation because it contains the = symbol.

equidistant The exact same DISTANCE apart at every point of reference, as in PARALLEL LINES.

equilateral triangle A TRIANGLE in which all three sides are equal in length.
See also SECTION IV CHARTS AND TABLES.

equivalent equations EQUATIONs that have the same SOLUTION or solutions.

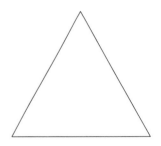

Equilateral triangle

Euler's f(x) notation The notation to indicate a FUNCTION, using "f" as the symbol for function, and *(x)* as the VARIABLE of the function.

evaluating expressions Determining the VALUE of an EXPRESSION by substituting the appropriate numbers for the VARIABLEs and then calculating the EQUATION.

even number Any number that is exactly divisible by 2. Numbers that end in 0, 2, 4, 6, or 8 are even numbers.
See also SECTION IV CHARTS AND TABLES.

evolution The reverse of INVOLUTION; the process of finding the root of a quantity.

exponent Numbers or symbols used to identify the POWER to which an EXPRESSION is to be multiplied. For example, 5^2 is read "five to the second power" or "five squared" and means 5×5. "Five to the third power" or "five cubed" is written as 5^3, and means $5 \times 5 \times 5$. "Five to the nth power" is written as 5^n, and means that 5 is multiplied times itself an undetermined, or *nth,* number of times. If an exponent follows a term of more than one VARIABLE, it only raises the power of the variable immediately before it. For example, $3ab^2$ only raises b to the second power, xy^3 only raises y to the third power, and so on.

exponentiation (involution) The process of raising any quantity to a POWER, or of finding the power of a number.

expression (algebraic expression) A grouping of one or more TERMS, which contains at least one number or VARIABLE, and includes addition, subtraction, MULTIPLICATION, or division. Each of the following is an expression: $3x$, $2 + 5y$, $7xy - 3x + y^2$. The first expression, $3x$, contains one term and is also called a MONOMIAL. The second expression, $2 + 5y$, contains two terms and is also called a BINOMIAL. The third expression, $7xy - 3x + y^2$, contains three terms and is also called a TRINOMIAL. Any expression with four or more terms is simply called a POLYNOMIAL. Binomials and trinomials are specific kinds of polynomials. Expressions do not contain an equals sign.
 See also SECTION IV CHARTS AND TABLES.

Extended Distributive Property The property that dictates how to multiply two POLYNOMIALS, in which each TERM in the first polynomial is multiplied by each term in the second polynomial.

f(x) *See* EULER'S f(x) NOTATION.

factor Any number or quantity that divides another number without leaving a REMAINDER.

factoring Dividing a number by its factors.

Fibonacci sequence A SEQUENCE of numbers beginning with 1, 1, in which each following number is the SUM of the two numbers immediately preceding it. The sequence starts 1, 1, 2, 3, 5, 8, 13, 21, 34, 55, etc . . . , and can be calculated using a RECURSIVE FORMULA. The Fibonacci sequence is named for Italian mathematician LEONARDO FIBONACCI.
 See also SECTION IV CHARTS AND TABLES.

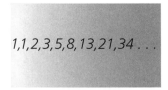

1,1,2,3,5,8,13,21,34 . . .

Fibonacci sequence

finite set A SET in which all of the elements or MEMBERS of the set can be listed, including the first and last elements. For example $A = \{x \mid x =$ all EVEN NUMBERS between 0 and 10$\}$. Since all of the elements can be listed, because $A = \{2, 4, 6, 8\}$, this is a finite set.

first-degree equation Any EQUATION with only one VARIABLE that is not multiplied by itself, is not part of a DENOMINATOR, and is involved with addition, subtraction, MULTIPLICATION, and/or division.

first-degree inequality Any statement of INEQUALITY with only one VARIABLE that is not multiplied by itself, is not part of a DENOMINATOR, and is involved with addition, subtraction, MULTIPLICATION, and/or division.

F.O.I.L. A method used for multiplying two POLYNOMIALS. F.O.I.L. stands for the order in which the MULTIPLICATION is done: **F**irst, **O**utside, **I**nside, **L**ast.
See also SECTION IV CHARTS AND TABLES.

fraction Any EXPRESSION that is written as two quantities in which one is divided by the other. For example, $\frac{1}{2}, \frac{3}{4}, \frac{5}{8}, \frac{7}{16}, \frac{14}{x} + y$.

fractional equation Any EQUATION in which a DENOMINATOR contains a VARIABLE.

fractional exponents An EXPONENT that is in FRACTION form, the NUMERATOR of which denotes the POWER, and the DENOMINATOR of which denotes the root. The fractional exponent EQUATION is written as $x^{n/b} = (^b\sqrt{x})^n = {}^b\sqrt{x^n}$.
See also SECTION IV CHARTS AND TABLES.

function A SET of ORDERED PAIRS in which all of the x-coordinate numbers are different. For example, this set of ordered pairs is not a function because all of the x-coordinates are not different: (3, 5), (5, 2), (5, 6), (8, 3). This set of ordered pairs *is* a function because all x-coordinates are different: (2, 7), (4, 3), (7, 4), (9, 6).

Fundamental Theorem of Algebra Any POLYNOMIAL that has complex COEFFICIENTS has a minimum of one complex number SOLUTION.

graph 1. The SET of points on a CARTESIAN COORDINATE SYSTEM that indicates the SOLUTION to an EQUATION.
2. A chart that visually compares quantities.

greatest common divisor *See* GREATEST COMMON FACTOR.

greatest common factor (greatest common divisor, greatest common measure) The largest number that can be divided into two other numbers without leaving a REMAINDER. For example, COMMON FACTORS of the numbers 4 and 8 are: 1, 2, and 4. The largest number is the greatest common factor, in this case 4.

greatest common measure *See* GREATEST COMMON FACTOR.

heptagon A POLYGON with seven sides and seven interior ANGLES.
See also SECTION IV CHARTS AND TABLES.

hexagon A six-sided POLYGON.
See also SECTION IV CHARTS AND TABLES.

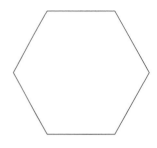

Hexagon

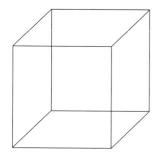

Hexahedron

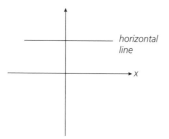

Horizontal line

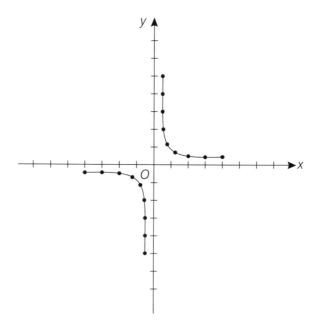

Hyperbola

hexahedron A SOLID three-dimensional HEXAGON, with six equal SQUARE sides; a CUBE.

highest common factor The largest number that can be factored out of both the NUMERATOR and the DENOMINATOR.

　　See also GREATEST COMMON FACTOR; SECTION IV CHARTS AND TABLES.

histogram A BAR GRAPH that shows both the type and frequency of a value.

horizontal axis The x-axis in a CARTESIAN COORDINATE SYSTEM graph.

horizontal line A line on a GRAPH that has no SLOPE and intersects the y-axis at only one point.

hyperbola The shape of a DISCONTINUOUS GRAPH of an inverse variation in algebra.

hypotenuse The side of a RIGHT TRIANGLE that is opposite the RIGHT ANGLE. This is also always the longest side of a right triangle.

　　See also SECTION IV CHARTS AND TABLES.

hypothesis A theory that is not proved, but is supposed to be true, so that it can be further tested for PROOF.

icosahedron A 20-sided POLYHEDRON, in which all 20 faces are EQUILATERAL TRIANGLES.
> *See also* SECTION IV CHARTS AND TABLES.

identity Any EQUATION in which all real values of the VARIABLE will make the equation true.

Identity Property Any quantity or number is equal to itself, written as $a = a$.
> *See also* SECTION IV CHARTS AND TABLES.

Identity Property of Addition *See* ADDITIVE IDENTITY PROPERTY.

Identity Property of Multiplication *See* MULTIPLICATIVE IDENTITY PROPERTY.

imaginary number The SQUARE ROOT of any negative REAL NUMBER.

imperfect number Any number whose FACTORS (excluding the number itself), when added up, equal a SUM more than or less than the number itself. Both ABUNDANT NUMBERS and DEFICIENT NUMBERS are imperfect numbers.

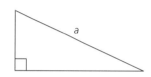

Hypotenuse

Icosahedron

improper fraction Any FRACTION in which the NUMERATOR is greater than or equal to the DENOMINATOR. For example, $\frac{3}{2}, \frac{7}{3}, \frac{5}{4}, \frac{6}{6}, \frac{12}{8}, \frac{20}{16}$, and so on.

See also SECTION IV CHARTS AND TABLES.

inconsistent equations EQUATIONS that have no SOLUTIONS.

independent variable The VARIABLE in a direct variation formula that does not rely on the other variable for its value. For example, in $y = kx^n$, y is the DEPENDENT VARIABLE, and x is the independent variable.

index A number or variable used in ROOTS and POWERS to indicate the power of a quantity. For example, when taking a root of a number, as in $\sqrt[3]{9}$ the index is 3; if no root is indicated, as in $\sqrt{25}$, the index is 2 (for the SQUARE ROOT). When raising a number to a power, as in 4^2, the index is 2; in 4^3, the index is 3, and so on, and in this case the index is often called the EXPONENT or the power.

inequality A MATHEMATICAL SENTENCE that contains quantities that are not equal, or might not be equal, in VALUE. For example, $3x + 6 < 29$, or $y - 8 \geq 45$.

inequality sign Any of the symbols used to show that two quantities are not equal or might not be equal to each other. For example: $<$, $\leq$, $>$, $\geq$, $\neq$. The largest side of the inequality sign always opens to the largest number, and the smallest side always points to the smallest number.

See also SECTION IV CHARTS AND TABLES.

infinite set A SET in which all of the elements, or MEMBERS, of the set cannot be listed, including the first and last elements. For example, $A = \{x \mid x = \text{all EVEN NUMBERS} > 0\}$.

integer Any WHOLE NUMBER, whether negative, positive, or zero. For example, $-5, -4, -3, -2, -1, 0, 1, 2, 3, 4, 5 \ldots$ are all integers.

See also SECTION IV CHARTS AND TABLES.

intercept The point on a Cartesian GRAPH where a line crosses an axis. If the line crosses the x-axis it is called the x-intercept, and if crosses the y-axis it is called the y-intercept.

interdependent event An event that can only occur if another event occurs. The PROBABILITY of two or more interdependent events occurring is calculated as the PRODUCT of the probabilities of each event occurring.

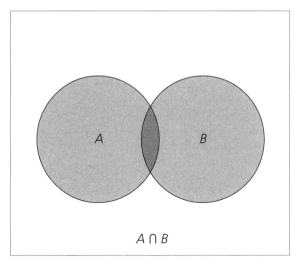

$A \cap B$

intersection A solution SET that includes only the elements that are common to each individual set. For example, if A = {all EVEN NUMBERS between 0 and 10} and B = {all numbers between 7 and 15}, then the intersection of A and B includes only the number 8, because 8 is the only element that both sets have in common. Intersection is identified with the symbol $\cap$. This example is written $A \cap B = 8$.

inverse operations A method of undoing one operation with another. For example, addition and subtraction are inverse operations of each other, and MULTIPLICATION and division are inverse operations of one another. Inverse operations are often used to check answers for accuracy, for example, using addition to check subtraction, and using multiplication to check division.
See also SECTION IV CHARTS AND TABLES.

involution *See* EXPONENTIATION.

irrational number Any REAL NUMBER that cannot be written as a SIMPLE FRACTION; any number that is not a RATIONAL NUMBER. For example, pi and the SQUARE ROOT of 2 are both irrational numbers.
See also SECTION IV CHARTS AND TABLES.

isosceles right triangle An ISOSCELES TRIANGLE that, in addition to having two equal sides, has one ANGLE that measures exactly 90°.
See also SECTION IV CHARTS AND TABLES.

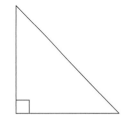

Isosceles right triangle

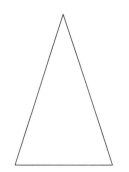

Isosceles triangle

isosceles trapezoid A QUADRILATERAL with two nonparallel sides that are equal in length.
> *See also* SECTION IV CHARTS AND TABLES.

isosceles triangle A TRIANGLE with two sides that are equal in length.
> *See also* SECTION IV CHARTS AND TABLES.

Law of Fractions When the NUMERATOR and DENOMINATOR are divided by the same number or multiplied by the same number, the value of the FRACTION remains the same. For example, if the numerator and denominator in the fraction $\frac{4}{8}$ are both multiplied by 2, the fraction becomes $\frac{8}{16}$, and is still the same value as $\frac{4}{8}$. If the numerator and denominator are both divided by 2, the fraction becomes $\frac{2}{4}$, and the value remains the same. This law is used for finding the lowest and highest common TERMS of a fraction.

leading coefficient The COEFFICIENT with the highest degree in a POLYNOMIAL.

least common denominator The number that is the LEAST COMMON MULTIPLE in the DENOMINATOR of two or more FRACTIONS. For example, in the fractions $\frac{3}{4} + \frac{2}{3}$, the least common denominator is 12, as it is the lowest common MULTIPLE of the denominators of 4 and 3, and when factored the fractions would become $\frac{9}{12} + \frac{8}{12}$. In the fractions $\frac{2}{3} + \frac{3}{6} + \frac{7}{9}$, the least common denominator is 18 and the fractions would become $\frac{12}{18} + \frac{9}{18} + \frac{14}{18}$.

least common multiple The number that is smallest in value of common MULTIPLEs. For example, 12, 24, and 36 are common multiples of both 3 and 4, and the least common multiple is 12.

legs The two sides of a RIGHT TRIANGLE that are not the HYPOTENUSE.

like terms Any TERMS that have the same VARIABLE but different COEFFICIENTS. For example, $3x + 7x$ has the same variable x.

line segment A piece of a straight line that is a specific length, and usually marked at the ends, for example, with A and B, to show that it is a particular line segment. For example, A_____B. The shortest DISTANCE between two points is a straight line, which is a line segment.

linear equation An EQUATION that has two CONSTANTS and one VARIABLE, often written as $ax + b = 0$. An equation with exactly two variables, each of which is involved with addition, subtraction, MULTIPLICATION, or division, and neither of which is raised to a POWER above 1, nor in a DENOMINATOR, nor is the PRODUCT of the two TERMS, is called a linear

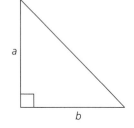

Legs

equation in two variables. For example, $x - y = 3$, $x + 2y = 14$, $4x - 9 = \frac{y}{5}$, etc. This is often written as $ax + by + c = 0$. If the equation has the same criteria and three variables, it is called a linear equation in three variables, and is often written as $ax + by + cz + d = 0$.

linear sequence *See* ARITHMETIC SEQUENCE.

list method of specification *See* ROSTER METHOD OF SPECIFICATION.

literal equation Any EQUATION that contains letters, VARIABLES, and numbers. For example, $4x + 2y = ax + y$.

logarithm An EXPONENT that represents a number, based on a system that uses a common base and its exponents to represent number values. For example, in a base 10 system, $10,000 = 10^4$. The exponent becomes the log that represents the number, so 4 is the log of 10,000. This is useful for dealing with very large numbers and the RATIOS between them on a scale, such as the Richter scale.

logarithmic scale A scale that spaces the DISTANCES between quantities as RATIOS, for example, the Richter scale, as opposed to the linear scale, which spaces the distances equally between units.

lowest terms The canceling of all COMMON FACTORS in both the NUMERATOR and DENOMINATOR. For example, the FRACTION $\frac{15}{25}$ is reduced to its lowest terms of $\frac{3}{5}$ by canceling out the common FACTOR of 5 in both the numerator and the denominator. In finding the lowest terms, the fraction is divided by the HIGHEST COMMON FACTOR.
  See also SECTION IV CHARTS AND TABLES.

mathematical sentence Any mathematical phrase that includes any of the following symbols: $<$, $\leq$, $>$, $\geq$, $=$, $\neq$. For example, $3x + y \leq 15$, $7 < x < 12$, $-3 \times -3 = 9$ are each a mathematical sentence. Any mathematical sentence with an equals sign, such as $-3 \times -3 = 9$, is specifically called an EQUATION.

matrix A rectangular chart of rows and columns, used to compare quantities or data.

mean (average) The value obtained from the SUM of a SET of numbers, divided by the amount of numbers in that set. For example, in the set of 2, 5, 6, 8, 9, 15 the mean is obtained from the sum (45) divided by the amount of numbers in the set (6), so the mean is 7.5.

median The middle value in an ordered SET of values. For example, in the set of 8, 12, 19, 22, 35, the median is 19. In an even-numbered set, the median is the AVERAGE of the middle two values. For example, in the

16
89
134 ← median
147
162

Median

set of 26, 34, 38, 45, the median is 36, which is the average of the middle two values: $(34 + 38) \div 2 = 36$.

members (elements) The individual components of a SET. When listed, the members of a set are usually enclosed within braces { }. For example, $A = \{2, 4, 6, 8\}$. To indicate that one of these elements is a member of the set A, we use the symbol $\in$, which is read "is a member of" or "belongs to." So, $2 \in A$, is read 2 is a member of A.

metric system The international DECIMAL system of weights and measurements that was developed in France, using the units of second for time, meter for length, and kilogram for weight.

minuend The number or quantity from which another number or quantity is subtracted. For example, in $365 - 14$, the minuend is 365, in $14y - x$, the minuend is $14y$.
See also SUBTRAHEND.

mixed number Any number consisting of both an INTEGER and a FRACTION or DECIMAL. For example, $3\frac{1}{2}$ and 3.5 are both mixed numbers.

monomial Any EXPRESSION that consists of just one TERM. For example, $2x$, $3xy^2$, or $4x^3y^2$. Expressions with more than one term are types of POLYNOMIALS.
See also SECTION IV CHARTS AND TABLES.

multiple Any number that is divisible by another number without leaving a REMAINDER. For example, 3 is a factor of 6, 9, 12, 15, 18, 21, 24. Each of these numbers is a multiple of 3.

multiplicand Any number that is multiplying an original number. For example, in 3×4, the multiplicand is 4.
See also MULTIPLICATION; MULTIPLIER.

multiplication The process, for POSITIVE NUMBERS, of adding a number to itself a certain number of times. For example, 3×4 is the same as adding $3 + 3 + 3 + 3$. There are certain rules that apply when multiplying numbers other than positive numbers, for example, multiplying by zero, multiplying a negative and a positive, or multiplying a negative and a negative.

Multiplicative Identity Property The PRODUCT of any number multiplied by 1 is the number itself, and the SUM of any number added to zero is the number itself. For example, $3 \times 1 = 3$.
See also SECTION IV CHARTS AND TABLES.

Multiplicative Property of Equality If two REAL NUMBERs are equal in value, for example, if $a = b$, then when they are multiplied by another real number, for example, c, the SUMs will be equal. In EQUATION form, it looks like this: if $a = b$, then $ac = bc$.
See also SECTION IV CHARTS AND TABLES.

Multiplicative Property of Zero If any REAL NUMBER is multiplied by zero, the sum will equal zero.
See also SECTION IV CHARTS AND TABLES.

multiplier Any number that is being multiplied. For example, in 3×4, 3 is the multiplier.
See also MULTIPLICATION; MULTIPLICAND.

natural number Beginning with the number 1, any positive, WHOLE NUMBER. For example, 3 is a natural number, –3 is not a natural number, and 0.3 is not a natural number. Natural numbers are also called *counting numbers*.
See also SECTION IV CHARTS AND TABLES.

negative number Any number that is less than zero in value. A negative number is indicated with the "–" sign.

nonagon A nine-sided POLYGON.

null set *See* EMPTY SET.

number The word *number* is typically used to mean "REAL NUMBER."

number line A straight line that represents all of the numbers. It is drawn with an arrow on the ends to show that it goes on indefinitely. Zero on the number line is called the ORIGIN, with NEGATIVE NUMBERs marked on points to the left of zero, and POSITIVE NUMBERs marked on points to the right of zero.

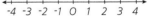

Number line

numerator (divisor) The number in a FRACTION that is above the division line. The number that divides the DIVIDEND. For example, in $\frac{1}{2}$, the numerator is 1. In $\frac{3}{4}$ the numerator is 3, in $\frac{7}{8}$ the numerator is 7, in $\frac{9}{16}$ the numerator is 9, and in $\frac{97}{100}$ the numerator is 97.
See also SECTION IV CHARTS AND TABLES.

numerical value *See* ABSOLUTE VALUE.

oblique angle Any ANGLE, whether acute or obtuse, that is not a RIGHT ANGLE.

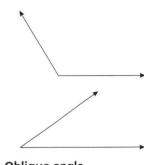

Oblique angle

Oblique-angled triangle

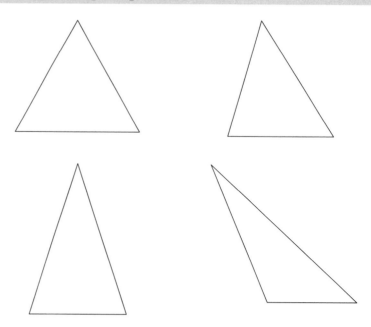

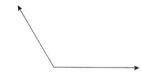

Obtuse angle

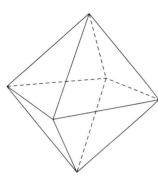

Octahedron

oblique-angled triangle A TRIANGLE that has no RIGHT ANGLES.
See also SECTION IV CHARTS AND TABLES.

oblique lines Any two lines that meet to form an OBLIQUE ANGLE.

obtuse angle Any ANGLE that measures between 90° and 180°.

octahedron An eight-sided POLYHEDRON, in which all eight faces are
EQUILATERAL TRIANGLES.
See also SECTION IV CHARTS AND TABLES.

odd number Any NUMBER that is not exactly divisible by 2.
See also SECTION IV CHARTS AND TABLES.

odds The RATIO of an event occurring, based on the number of ways it can
occur divided by the number of ways it will not occur.

open sentence Any MATHEMATICAL SENTENCE that contains a VARIABLE, and
which may be true or false depending on the value SUBSTITUTEd for
the variable.

operations The practice of doing addition, subtraction, MULTIPLICATION, or
division. Some of these operations undo each other, and are called

INVERSE OPERATIONS, and in some instances the order in which these operations are performed is important in solving EQUATIONS and evaluating expressions.

See also ORDER OF OPERATIONS.

ordered pairs A pair of numbers written as *(x, y),* in which the order is significant. For a GRAPH, the first number corresponds with the point on the *x*-axis, and the second number corresponds with the point on the *y*-axis. An ordered pair of (3,7) would not graph to the same point as an ordered pair of (7, 3).

order of operations The order for doing math when EVALUATING EXPRESSIONS, abbreviated as P.E.M.D.A.S.

See also SECTION IV CHARTS AND TABLES.

ordinate On an *(x, y)* graph, the *x* coordinate is the ABSCISSA, and the *y* coordinate is the *ordinate.* Together, the abscissa and the ordinate make the coordinates.

origin The INTERSECTION point of the *X*-AXIS and the *Y*-AXIS on a graph, indicated with *O.* Its ordered pair is (0, 0).

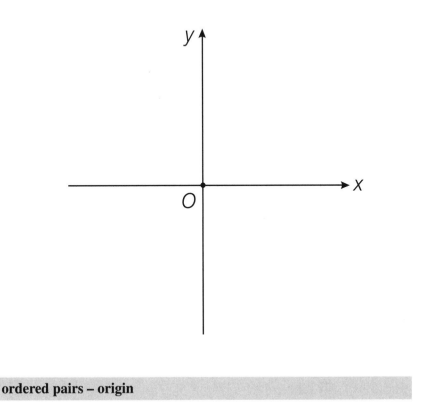

Origin

Parabola

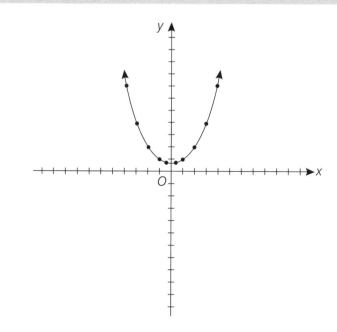

parabola The curved line GRAPH formed by plotting a QUADRATIC
EQUATION. The parabola contains a VERTEX, an axis that runs
through the vertex, called the axis of symmetry, around which the
points of the curve of the parabola on either side of the axis are
symmetrical, and has x- and y-intercepts. The parabola curves to
point upward if the COEFFICIENT of the SQUARED term is positive,
and downward if it is negative.

paraboloid The SOLID of a PARABOLA.

parallel lines Two or more lines that are exactly the same DISTANCE apart.
Parallel lines cannot intersect.

Parallel lines

parallelogram A four-sided PLANE figure with opposite sides that are parallel
and equal in length.

Parallelogram

parallel planes Two or more PLANES that are exactly the same DISTANCE apart. Parallel planes cannot intersect.

Pascal's triangle Originally used in India, and again in the mid-1500s, this triangle was named after French mathematician BLAISE PASCAL, who discovered new properties in the triangle he called *triangle arithmetique.* This is a two-dimensional representation of an ARITHMETIC SEQUENCE, beginning with 1 in the top row (called Row 0), and 1 in the sides of the triangle, in which the numbers next to each other in the triangle equal the number below it.

    See also SECTION IV CHARTS AND TABLES.

P.E.M.D.A.S. The ORDER OF OPERATIONS used to evaluate an EXPRESSION. P.E.M.D.A.S. stands for **P**arentheses, **E**xponents, **M**ultiplication, **D**ivision, **A**ddition, **S**ubtraction.

    See also SECTION IV CHARTS AND TABLES.

pentagon A POLYGON with five sides and five interior ANGLES.

perfect number Any NUMBER whose FACTORS (excluding the number itself), when added up, equal *exactly* the number itself. For example, the factors for the number 28 are 1, 2, 4, 7, and 14. When these numbers are added, the SUM is 28, making this a perfect number.

perfect square Any WHOLE NUMBER that is the exact SQUARE of another whole number. For example, 2 is the exact square of 4, so 2 is a perfect square. Other perfect squares include 1 (which is $\sqrt{1}$), 3 ($\sqrt{9}$), 4 ($\sqrt{16}$), 5 ($\sqrt{25}$), 6 ($\sqrt{36}$), and so on.

perimeter The DISTANCE around a POLYGON. The length of the perimeter is found by adding up the length of each of the sides.

perpendicular The relationship of two lines that come together to form a RIGHT ANGLE. The angle is a right angle, and the lines are perpendicular.

pi (π) The RATIO of the CIRCUMFERENCE of a CIRCLE divided by the DIAMETER. The approximation of this IRRATIONAL NUMBER is usually written as 3.1416.

plane A flat, straight, two-dimensional surface, which can be either real or imaginary. An example of a plane with finite edges is a blackboard, a tabletop, or a side of a CUBE. Some imaginary planes can be infinite

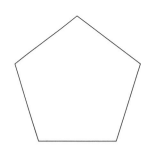

```
                  1
               1     1
            1     2     1
         1     3     3     1
      1     4     6     4     1
   1     5    10    10     5     1
1     6    15    20    15     6     1
1  7   21    35    35    21    7    1
```

Pascal's triangle

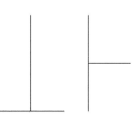

Pentagon

Perpendicular

Pi

in size, such as the plane represented in a CARTESIAN COORDINATE SYSTEM graph.

point   The COORDINATES of an ordered pair on a Cartesian graph.

polygon   Any closed figure in a PLANE that has three or more LINE SEGMENTS. A TRIANGLE is a three-sided polygon; a SQUARE and a RECTANGLE are each four-sided polygons.
> *See also* REGULAR POLYGON; SECTION IV CHARTS AND TABLES.

Polygon

polyhedron   A three-dimensional figure in which all of the faces of the figure are POLYGONS.
> *See also* SECTION IV CHARTS AND TABLES.

polynomial   Any EXPRESSION that consists of a string of MONOMIALS. For example, $\frac{1}{2}x + 5xy^2 + \pi$. Polynomials with just two TERMS are called BINOMIALS, and polynomials with exactly three terms are called TRINOMIALS.
> *See also* SECTION IV CHARTS AND TABLES.

positive number   A number that is greater than zero in value. A positive number that is written as a SIGNED NUMBER is written with a "+" sign. If a number is written without a sign, it is considered to be a positive number.

postulate   *See* AXIOM.

power   Usually refers to the EXPONENT, and indicates the number of times a particular quantity is multiplied by itself. For example, 3 to the power of 2, written as 3^2, indicates that 3 is multiplied by itself 2 times (3×3); 10 to the power of 3, written as 10^3, means that 10 is to be multiplied by itself 3 times $(10 \times 10 \times 10)$; 5 to the power of 5, written as 5^5, shows that 5 is to be multiplied by itself 5 times $(5 \times 5 \times 5 \times 5 \times 5)$.

powers of 10   Numbers with a base of 10, such as 10, 100, 1,000, 10,000, and so on, which can be written with an EXPONENT as 10, 10^2, 10^3, 10^4, and so on.

Powers Property of Equality   Equal numbers with equal powers are equal in value, written, if $a = b$, then $a^n = b^n$.

prime number   Any NUMBER that is divisible only by itself and 1. For example, 2, 3, 5, 7, 11, 13, 17, 19, 23, 29, and so on.
> *See also* SECTION IV CHARTS AND TABLES.

probability The RATIO that tells the likelihood of an event happening. If an event is certain to happen, the probability is 1. If it is impossible, the probability is zero. All probabilities are a ratio of zero, one, or somewhere in between. The calculation for probability is the total number of ways an event can occur, divided by the total possible number of events.

product The total obtained by multiplying numbers or quantities together. For example, in $6 \times 3 = 18$, the total 18 is the product; in $1 \times 5 \times 9 = 45$, the total of 45 is the product, and in $a \times b = ab$, the total ab is the product.

proof A series of statements using properties, definitions, and THEOREMs to show that a mathematical statement is true because all of the components of it are true.

proper fraction Any FRACTION in which the NUMERATOR is smaller than the DENOMINATOR. For example, $\frac{1}{2}, \frac{2}{3}, \frac{3}{4}, \frac{7}{8}, \frac{5}{16}$, and so on.
      See also SECTION IV CHARTS AND TABLES.

proper subset A proper subset is a SET in which all of the elements of one set are part of another set, and the other set has additional elements that do not belong to the SUBSET. For example, if $A = \{2, 4\}$, and $B = \{2, 4, 6, 8\}$, then A is a proper subset of B, because B has additional elements in its set that are not included in A. This is written as $A \subset B$.

pyramid A three-dimensional polyhedral with a polygonal base and triangular faces. A typical pyramid base might be a TRIANGLE or a SQUARE, and

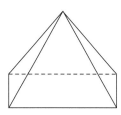

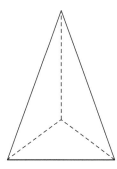

Pyramid

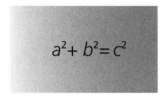

Pythagorean theorem

the faces of a pyramid always join at the top to form a common point, or VERTEX.

See also SECTION IV CHARTS AND TABLES.

Pythagorean theorem Used to determine the length of a side of a RIGHT TRIANGLE, it states that the SQUARE of the length of the HYPOTENUSE equals the SUM of the SQUARED lengths of the other two sides. This is written as $c^2 = a^2 + b^2$.

See also SECTION IV CHARTS AND TABLES.

quadrant Each of the four sections of a CARTESIAN COORDINATE SYSTEM GRAPH that are the result of the interception of the x-axis and y-axis.

quadratic equation Any EQUATION with only a SQUARED TERM as its highest term.

See also SECOND-DEGREE EQUATION.

quadratic equation standard form All TERMs of the QUADRATIC EQUATION are on the left side of the equals sign, and zero is on the right side, and the terms can be expressed as $ax^2 + bx + c = 0$. For example, $x^2 + 4x + 7 = 0$.

quadratic inequality Any MATHEMATICAL SENTENCE that is identical in form to a QUADRATIC EQUATION, except that it contains an INEQUALITY SIGN instead of an equals sign. The standard form of a quadratic inequality is the same as that of a quadratic equation, with the TERMS on the left side of the inequality sign, and zero on the right side, for example, $ax^2 + bx + c < 0$.

quadrilateral (tetragon) A POLYGON that has four sides and four ANGLES.

See also SECTION IV CHARTS AND TABLES.

quartic Fourth-degree term. For example, x^4.

quartic equation An EQUATION that contains a TERM of the fourth degree as its highest POWERED term. An equation containing an x^2 is a SECOND-DEGREE EQUATION, an x^3 is a THIRD-DEGREE EQUATION, an x^4 is a fourth-degree equation, x^5 is a fifth-degree equation, and so on.

quaternary Four.

quintic Fifth-degree term. For example, x^5.

quintic equation An EQUATION that contains a TERM of the fifth degree as its highest POWERED term. An equation containing an x^2 is a

SECOND-DEGREE EQUATION, an x^3 is a THIRD-DEGREE EQUATION, an x^4 is a fourth-degree equation, x^5 is a fifth-degree equation, and so on.

quotient The total obtained from dividing one number by another number. For example, in $14 \div 2 = 7$, the total 7 is the quotient.

radical (root sign) The name for the symbol used for taking the ROOT of a number, which looks like this: $\sqrt{}$.

radicand The number located under the RADICAL sign. For example, in $\sqrt{9}$, the radicand is 9; in $\sqrt{25}$, the radicand is 25, and in $\sqrt{36}$, the radicand is 36. This number is also sometimes called the base.

radii Two or more RADIUS.

radius A LINE SEGMENT that extends from the center of a CIRCLE to any point on the circle, or from the center of a REGULAR POLYGON to any VERTEX of the polygon.

radix (base) The base of a number system. For example, in the DECIMAL system, the radix, or base, is 10. In a binary system, the radix is 2.

ratio The relationship of one number to another, expressed as a QUOTIENT. For example, the ratio of 2 to 4 is written as 2:4, or $\frac{2}{4}$, which can be simplified to $\frac{1}{2}$. The two elements of the ratio, the first TERM and the second term, are called the ANTECEDENT and the CONSEQUENT, respectively.

rational number Any INTEGER, or RATIO of integers, that can be written as a FRACTION, as long as zero is not the DENOMINATOR. For example, $\frac{3}{4}$, 4, $\frac{8}{2}$, $\frac{1}{4}$, .36.

real number Any number, whether rational or irrational, that can be written with a DECIMAL and is not an IMAGINARY NUMBER.

reciprocal The mirror image of a FRACTION, in which the NUMERATOR and DENOMINATOR are inverted. For example, the reciprocal of $\frac{1}{3}$ is $\frac{3}{1}$, the reciprocal of $\frac{2}{5}$ is $\frac{5}{2}$, and so on.
See also SECTION IV CHARTS AND TABLES.

rectangle A right-angled PARALLELOGRAM (QUADRILATERAL).
See also SECTION IV CHARTS AND TABLES.

recurring decimal *See* REPEATING DECIMAL.

recursive formula (recursive definition) A SET of statements for a SEQUENCE of numbers that explains the first few terms of the

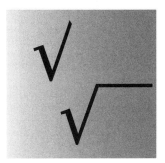

Radical

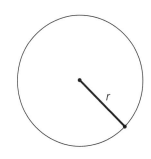

Radius

Rectangle

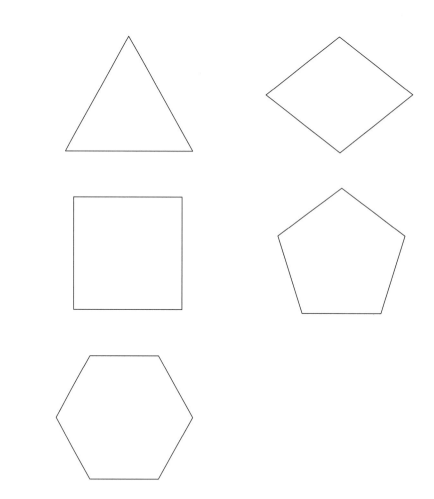

Reflex angle

sequence, plus the rule on how the *n*th TERM of the sequence is related to the first terms. An example of a recursive formula is the FIBONACCI SEQUENCE.

reflex angle Any ANGLE that measures greater than 180° and less than 360°.

Reflexive Property Any number is equal to itself, usually expressed as $a = a$. *See also* SECTION IV CHARTS AND TABLES.

regular polygon Any POLYGON in which all of the interior ANGLES are equal and all of the LINE SEGMENTS are equal.

Regular polygons

Repeating decimal

remainder The amount left over after one number is divided by another number. The DIFFERENCE is also sometimes called the remainder.

repeating decimal (recurring decimal) A DECIMAL FRACTION that has a pattern of numbers that repeats indefinitely. For example, 0.2956565656 . . . .

repetend The number or SET of numbers in a REPEATING DECIMAL that repeats. For example, in .0368951951951951 . . ., the repetend is 951.

replacement set (domain) A SET of numbers that can be used to replace VARIABLES in an EQUATION.

rhombus An oblique-angled PARALLELOGRAM with four equal sides. *See also* SECTION IV CHARTS AND TABLES.

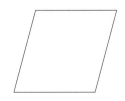

Rhombus

right angle An ANGLE that measures exactly 90°.

right triangle Any TRIANGLE containing a 90° ANGLE. *See also* SECTION IV CHARTS AND TABLES.

root Any number that is multiplied by itself a specific number of times to equal another number. For example, in $3 \times 3 \times 3$, the root is 3; in 4×4, the root is 4; in 5^4, the root is 5, and so on.

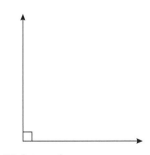

Right angle

roster method of specification (list method) The method in which the elements or MEMBERS of a SET are listed. For example, $A = \{2, 4, 6, 8\}$. An alternative method of listing elements of a set is the DESCRIPTION METHOD OF SPECIFICATION.

round angle An ANGLE that measures exactly 360°.

rounding down function The process of lowering a number to the next greatest number. For example, when rounding down the DECIMAL 4.312, the next greatest value is 4.31, and when rounding this decimal down to a WHOLE NUMBER, that next greatest value is 4. When rounding down the whole number 2,341 to the nearest hundred, that next greatest value is 2,300. The guide used to determine if a number should be rounded down or rounded up is that any number below 5 is rounded down, and any number greater than or equal to 5 is rounded up.

rounding up function The process of raising a number to the next greatest number. For example, when rounding up the DECIMAL 4.759, the next greatest value is 4.76, and when rounding this decimal up to a WHOLE NUMBER, that next greatest value is 5. When rounding up the whole

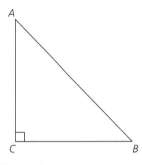

Right triangle

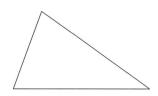

Round angle

number 2,978 to the nearest hundred, that next greatest value is 3,000. The guide used to determine if a number should be rounded down or rounded up is that any number below 5 is rounded down, and any number greater than or equal to 5 is rounded up.

rule method of specification *See* DESCRIPTION METHOD OF SPECIFICATION.

sampling Using a SUBSET of a population to draw conclusions about the characteristics of the overall population.

scale To change the DIMENSIONS of a figure horizontally, vertically, or both.

scalene triangle A TRIANGLE in which no two sides are equal in length. *See also* SECTION IV CHARTS AND TABLES.

Scalene triangle

scientific notation A number written as a FACTOR of any number between 1 and 10, multiplied by 10 with the corresponding EXPONENT. For example, the scientific notation for 100 is 1×10^2, for 1,264 is 1.264×10^3, and for 0.029 is 2.9×10^{-3}. The exponent is determined by how many places to the left or right the DECIMAL POINT moves to create a number between one and 10. 1,264 requires moving the decimal point three places to the left, giving the scientific notation of 1.264 (for a number between one and 10) $\times 10^3$ (for 10 with the corresponding exponent of 3).

secant Any line that dissects a CIRCLE at two points.

second-degree equation (quadratic equation) An EQUATION that contains a SQUARED TERM, which is also its highest POWERED term. For example, $x^2 = 16$, $x^2 + 5 = 14$, and $3x + 2x^2 - 9 = 3$, are all second-degree equations. An equation containing an x^3 is a THIRD-DEGREE EQUATION, an x^4 is a fourth-degree equation, x^5 is a fifth-degree equation, and so on.

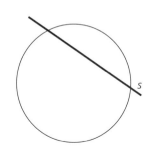

Secant

sense The direction in which an INEQUALITY SIGN points. If two signs point the same way, they have the same sense; for example, $3 < 5$ and $x < 7$. If they point in opposite directions, they have opposite sense; for example, $2 > x$ and $x < 7$.

sentence *See* MATHEMATICAL SENTENCE.

sequence Any ordered SET or list of quantities. The items in a sequence are called TERMS.

series The SUM of a specific SET of sequentially ordered TERMS.

set Any collection of numbers, objects, or elements. Sets are usually identified by capital letters. For example, if a set called A consists of the numbers 2, 4, 6, and 8, it is identified as $A = \{2, 4, 6, 8\}$.

sigma notation (summation notation) Derived from the 18th letter of the Greek alphabet, this symbol Σ is used to represent the SUM of all the values in an ARITHMETIC SEQUENCE.

signed number A number that is written with a plus sign (a POSITIVE NUMBER) or a minus sign (a NEGATIVE NUMBER). Both +26 and –14 are signed numbers because they are preceded by a plus sign and a minus sign, respectively.
    See also SECTION IV CHARTS AND TABLES.

similar figures Two or more figures that are exactly the same in shape, but not necessarily in size. (See illustration on page 38.)

simple fraction *See* COMMON FRACTION.

simplify Removing grouping symbols and COMBINING LIKE TERMS to bring the EQUATION or sentence to its simplest form.

simultaneous equations Any two EQUATIONS that, when solved together, are true at the same time.

slope The ANGLE of descent or ascent of a line. Slope is calculated as the rise, or vertical change, divided by the run, or horizontal change. For two COORDINATES, the slope is the DIFFERENCE in the y coordinates (the rise) divided by the difference in the x coordinates (the run), written as $y_2 - y_1/x_2 - x_1$. If the line slants downward, it is descending, or is said to have declivity, and the slope will be a NEGATIVE NUMBER. If the line slants upward, it is ascending, or has acclivity, and the slope will be a POSITIVE NUMBER. If the line has no slope, it is a HORIZONTAL LINE.

solid A three-dimensional shape, such as a CUBE, a SPHERE, or a HEXAHEDRON.
    See also SECTION IV CHARTS AND TABLES.

solution The number or quantity used in place of a VARIABLE that makes an EQUATION true.

solving the equation Finding the VALUE of the quantities in an EQUATION through algebraic computation.

Sigma notation

Similar figures

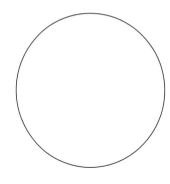

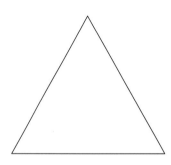
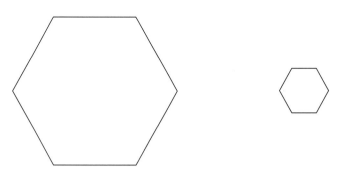

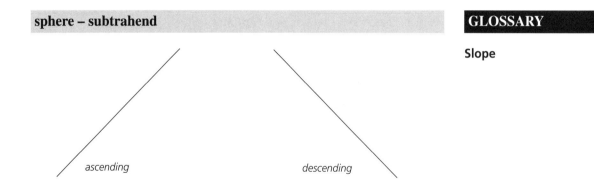

ascending   descending

Slope

sphere   A closed three-dimensional round SOLID that is, at all points, EQUIDISTANT from the center.
      See also SECTION IV CHARTS AND TABLES.

square   1. The second POWER of any number.
      2. A right-angled PARALLELOGRAM (QUADRILATERAL) in which all four sides are equal in length.
      See also SECTION IV CHARTS AND TABLES.

squared   A base number that is multiplied by itself, or raised to the second POWER. For example, 3 squared, written as 3^2, is 3×3; 5^2 is 5×5; 10^2 is 10×10, and so on.

square root   Any number that is one of two exactly equal FACTORS of another number. For example, $3 \times 3 = 9$, so the square root of 9 is 3; $5 \times 5 = 25$, so the square root of 25 is 5; $6 \times 6 = 36$, so the square root of 36 is 6, and so on. The symbol for square root, also called the RADICAL sign, looks like this: $\sqrt{}$.

straight angle   An ANGLE that measures exactly 180°.

subset   A subset is a SET in which all the elements of one set are included in the elements of another set. For example, if $A = \{2, 4\}$, and $B = \{2, 4, 6, 8\}$, then A is a subset of B. Also, if $A = \{2, 4, 6\}$, then A is a subset of B, or if $A = \{2, 4, 6, 8\}$, then A is a subset of B, because in every instance, all of the elements of A are included in the elements of B. This is written as $A \subseteq B$, and is read as A is contained in B.

substitute   To replace one value or quantity with another.

subtrahend   Any quantity or number that is subtracted from another quantity or number. For example, in $365 - 14$, the subtrahend is 14, and in $27y - x$, the subtrahend is x.
      See also MINUEND.

Sphere

Square

180°

Straight angle

sum The total obtained by adding numbers together. For example, in $7 + 3 = 10$, the total 10 is the sum; and in $3 + 5 + 9 = 17$, the total 17 is the sum.

summation notation *See* SIGMA NOTATION (Σ).

sum of two cubes A formula used to FACTOR two CUBED BINOMIALS into two POLYNOMIALS of the sum of the CUBE roots of the first TERM times the sum of the cube roots of the second, written as $x^3 + y^3 = (x + y)(x^3 - xy + y^3)$.

supplementary angles Two ANGLEs that, when summed, equal 180°.

Symmetric Property If two numbers have the same VALUE, they are symmetrical, or equal. This is usually expressed as: If $a = b$, then $b = a$. *See also* SECTION IV CHARTS AND TABLES.

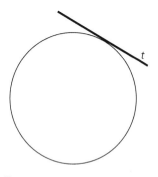

Tangent

tangent Any straight line that touches a curved line at only one point without intersecting.

term Any number, VARIABLE, or group of numbers and variables that form a MONOMIAL. For example, each of the following is a term: 2, $3x$, $4x^2y^3$. In FRACTIONS, the terms are the NUMERATOR and DENOMINATOR.

ternary Three.

tetragon *See* QUADRILATERAL; SECTION IV CHARTS AND TABLES.

tetrahedron A three-dimensional SOLID, a POLYHEDRON, in which all four faces are EQUILATERAL TRIANGLES. *See also* SECTION IV CHARTS AND TABLES.

theorem A statement that can be, or has been, proved.

third-degree equation An EQUATION that contains a CUBED term, and has no term POWERed higher than a cubed term. For example, $x^3 = 8$, $x^3 - x = 24$, and $3x + 2x^3 - 2 = 3$, are all third-degree equations. An equation containing an x^2 is a SECOND-DEGREE EQUATION; an x^4 is a fourth-degree equation, and so on.

Transitive Property of Equality Any numbers or quantities that are equal in value to the same quantity are also equal to each other. In EQUATION form, it looks like this: if $a = b$, and $b = c$, then $a = c$. *See also* SECTION IV CHARTS AND TABLES.

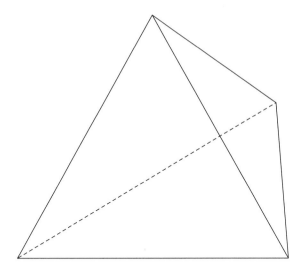

Tetrahedron

trapezium A four-sided POLYGON (QUADRILATERAL) that has no parallel sides.
See also SECTION IV CHARTS AND TABLES.

trapezoid A four-sided POLYGON (QUADRILATERAL) that has only two parallel sides.
See also SECTION IV CHARTS AND TABLES.

triangle A POLYGON that has three sides and three ANGLES. The SUM of the angles always equals 180°.
See also SECTION IV CHARTS AND TABLES.

trinomial A POLYNOMIAL that has exactly three TERMS. An EXPRESSION that consists of a string of three MONOMIALS. For example, $4x + 5y - 6x^2y$.
See also SECTION IV CHARTS AND TABLES.

unary One.

union A solution SET that includes all of the elements of each individual set, and all of the elements of the combined sets. For example, if $A = \{$all EVEN NUMBERS between 0 and 10$\}$ and $B = \{$all numbers between 7 and 15$\}$, then the union of A and B includes all of the even numbers between 0 and 10 and all numbers between 7 and 15. A union set is an inclusive set of all of the elements. Union is identified with the symbol $\cup$.

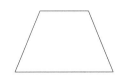

Trapezium

Trapezoid

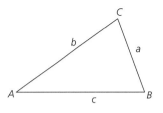

Triangle

Union

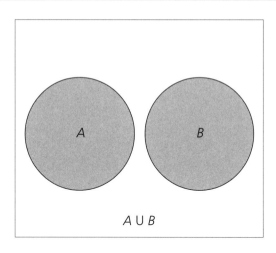

$A \cup B$

unit A single whole quantity; the number one.

value The numerical quantity of a VARIABLE or FUNCTION.

variable Letter or symbol used to represent unknown numbers or quantities. Some of the most common variables used in algebra are n, x, and y. Other variables often used are a, b, c, and z.
See also SECTION IV CHARTS AND TABLES.

velocity The rate at which an object travels through space and time.

Venn diagram A chart used to graphically show the relationship between SETS.

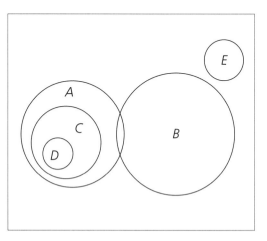

Venn diagram

GLOSSARY

unit – Venn diagram

Vertex

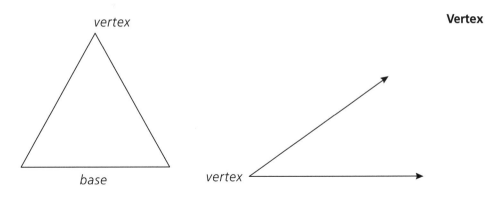

vertex

base

vertex

vertex 1. The common point from which two lines form an ANGLE.
2. The point on a TRIANGLE that is opposite the base of the triangle.

vertical angles ANGLEs on the opposite sides of a VERTEX, formed by two lines intersecting.

vertical axis The *y*-axis in a CARTESIAN COORDINATE SYSTEM graph.

vertical line A line on a GRAPH that has no SLOPE and intersects the *x*-axis at only one point.

volume The amount of space that a three-dimensional SOLID occupies. For example, the solid space of a CUBE, a cylinder, or a SPHERE is the volume. Volume is measured in CUBIC UNITS.
See also SECTION IV CHARTS AND TABLES.

whole number Any number that is not a FRACTION, beginning with zero.
See also SECTION IV CHARTS AND TABLES.

x 1. A common VARIABLE used in algebra to signify an unknown quantity.
2. The variable used to signify the HORIZONTAL LINE in a Cartesian GRAPH (*x*-axis).

x-axis The HORIZONTAL AXIS (horizontal NUMBER LINE) in a CARTESIAN COORDINATE SYSTEM graph. POSITIVE NUMBERS are to the right of the ORIGIN on the *x*-axis, and NEGATIVE NUMBERS are to the left of the origin on the *x*-axis.

x-intercept The point on a GRAPH in which a LINE SEGMENT intersects the *x*-axis, which is when $y = 0$.

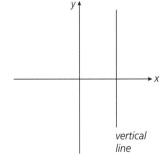

Vertical line

y-axis The VERTICAL AXIS (vertical NUMBER LINE) in a CARTESIAN COORDINATE SYSTEM graph. POSITIVE NUMBERS are up from the ORIGIN on the *y*-axis, and NEGATIVE NUMBERS are down from the origin on the *y*-axis.

y-intercept The point on a GRAPH in which a LINE SEGMENT intersects the *y*-axis, which is when $x = 0$.

z-axis The axis in a three-dimensional coordinate system that is PERPENDICULAR to the *xy*-plane.

z-coordinate The coordinate that corresponds to the *z*-axis in a three-dimensional coordinate system, listed as the third coordinate in an ordered triple as *(x, y, z)*.

zero The mathematical symbol used to show no measurable VALUE.

Zero, Multiplicative Property of *See* MULTIPLICATIVE PROPERTY OF ZERO.

Zero Exponent Theorem Any real non-zero number raised to the POWER of zero equals one. Written as: $b^0 = 1$.

Zero Product Property *See* MULTIPLICATIVE PROPERTY OF ZERO.

SECTION TWO
BIOGRAPHIES

Abel, Niels Henrik (1802–29) Norwegian who, in the words of
mathematician Charles Hermite, left the mathematical world
"enough to keep them busy for 500 years," at the time of his
death at the age of 27. Abel is most known for his work on
elliptic functions, and for his PROOF of a quintic EQUATION, in
which he shows that, unlike POLYNOMIALs of lower degrees,
any polynomial of the fifth degree (or higher) cannot be solved
in terms of radicals.

Abu Ali al-Hasan ibn al-Haytham (al-Basri, al-Misri, Alhazen)
(ca. 965–1040) Arab mathematician who made great
contributions to the field of optics, and worked in mathematics on
geometry, astronomy, and number theory. He purportedly wrote
more than 90 works, of which more than 55 still exist, including
an autobiography. Much of his scientific work supposedly
occurred when he pretended to be insane, so he would be
confined to his house and able to avoid the Caliph, whom he felt
was potentially dangerous. His work on number theory includes
some of the earliest examples of congruence theories. He became
interested in science after reading the work of Aristotle.

Agnesi, Maria Gaëtana (1718–99) Italian scholar who learned
several languages, including Latin, Greek, and Hebrew, by the
age of nine, and at the age of 20 began a project combining her
knowledge of language with her devotion to math. Ten years
later she published a two-volume work, *Analytical Institutions
(Instituzioni analitiche ad uso della gioventù italiana),* which
was valued as the most precisely ordered and largest
compilation of the work of the world's greatest mathematicians.
One piece in the book involved the discussion of a cubic curve
called *la versoria* (meaning "rope that turns a sail"). In an
1801 publication, this was mistakenly translated into English
as *l'aversiera,* which means "witch," and has ever since been
known as the "Witch of Agnesi."

Alembert, Jean Le Rond d' (1717–83) French mathematician, and
later philosopher, named after the church where his mother left
him as an illegitimate newborn child, d'Alembert spent a
lifetime surrounded by controversy for his argumentative
behavior and self-righteousness. He used his attitude to argue
his mathematical ideas, including improving on SIR ISAAC
NEWTON regarding the conservation of kinetic energy, and his

principle of mechanics, which he believed to be part of the mathematical sciences. He is most known for his work on the equilibrium and motion of fluids entitled *Traité de l'equilibre et du mouvement des fluides,* his article on vibrating strings in which he was the first to publish the wave equation (even though it was flawed), his pioneering work on partial differential EQUATIONS in physics, and his extensive scientific writings on philosophy and mathematics, most notably PROBABILITY and geometry, for the 28-volume *Encyclopédie,* as well as d'Alembert's RATIO test.

Apollonius of Perga (ca. 262–190 B.C.E.) The Greek mathematician often called "the Great Geometer," whose most famous work is his text on conics, also worked with TANGENTS, RATIOS, and inclinations, among others. He coined such terms as PARABOLA, HYPERBOLA, and ellipse, and gave an approximation for pi as 3.1416, although he never explained how he arrived at this VALUE.

Argand, Jean-Robert (1768–1822) A self-taught amateur mathematician, this Swiss native lived in Paris and worked on creating his most renowned work, graphing COMPLEX NUMBERS of the form $a + bi,$ using one AXIS to represent IMAGINARY NUMBERS, and the other to represent REAL NUMBERS, which is now known as the Argand diagram. His book on this subject, *Essai sur une manière de représenter les quantitiés imaginaires dans les constructions géométriques,* was originally self published, and his authorship was unknown for seven years. He shares credit for originating this idea with Caspar Wessel, whose paper on the same topic was published in 1799 but went unnoticed by the mathematical community. Argand's impressive body of work also included a PROOF on the FUNDAMENTAL THEOREM OF ALGEBRA, in which he was the first person to use complex numbers for the COEFFICIENTS.

Archimedes of Syracuse (ca. 287–12 B.C.E.) Ancient Greek considered with SIR ISAAC NEWTON and KARL FRIEDRICH GAUSS to be one of the three greatest mathematicians of all time, he is best known for his geometry and his mathematical-intensive inventions, such as the catapults used to protect Syracuse during a three-year siege, and the Archimedean screw, a corkscrew-like device used to move water out of flooded fields or the hold of a ship. Archimedes devised propositions to determine AREAS (CIRCLES, PARABOLA) and RATIOS, and worked on PARALLEL LINES, TRIANGLES, SPHERES, and cylinders. While taking a bath, he formulated the principle of buoyancy, and the story goes that in his excitement he jumped up and ran naked through the streets of Syracuse, shouting "Eureka!" over his discovery.

Archytas of Tarentum (ca. 428–350 B.C.E.) This Greek mathematician was a follower of PYTHAGORAS OF SAMOS, and believed that mathematics was the foundation for everything. He studied the four branches of mathematics—geometry, arithmetic, astronomy, and music—and his work was the basis of Plato's writings on sound theory. He found a SOLUTION to the Delian Problem of duplicating the CUBE, and stated that there is no geometric MEAN between two numbers in the RATIO $n + 1 : n,$ which is agreed to be the foundation of EUCLID OF ALEXANDRIA's work in his *Elements* Book VIII.

Aristarchus of Samos (ca. 310–230 B.C.E.) Called "Aristarchus the mathematician" by the Greeks, he is more commonly known for his work in astronomy, and is sometimes called "the Ancient Copernicus." But his observations of astronomy were based on his original mathematical calculations. Aristarchus' diagram showing the triangular relationship between the Sun, the Earth, and the Moon, was an attempt to calculate the DISTANCES based on RATIOS (geometry and trigonometry). He invented a bowl-shaped sundial to more accurately tell time, and through his computations determined that the Sun, not the Earth, was the center of the universe. Despite the mathematical soundness, this was of course contrary to Aristotle's teachings, and did not catch on until Copernicus came along.

Aryabhata II (ca. 920–1000) Like most Indian mathematicians of this period, much of Aryabhata II's computations were dedicated to astronomy. His extensive work *Mahasiddhanta* also covered algebra and geometry, specifically giving rules for solving the indeterminate EQUATION $by = ax + c,$ and calculating CUBE ROOTS.

Babbage, Charles (1791–1871) This British mathematician invented a small mechanical calculator in 1812. He published a table of

LOGARITHMS from 1 to 108000, invented a cowcatcher for the front of a train, invented colored lights for the theater that were never used in public because of fear that they would cause a fire, was lowered into a volcano to observe lava, and created uniform postal rates. His greatest accomplishment was also his biggest defeat during his lifetime. In 1823 he started work on the "Difference Engine," the world's first computer. Politics, public ridicule, and lack of enough funding to finish the work caused the project to drag out until 1842, when the British government decided that, after not funding the project for the past eight years, it would officially stop funding. Babbage's work on his Difference Engine, and his drawings on the Analytical Engine, a larger computerized machine, were the basis for modern computing.

Bari, Nina Karlovna (1901–61) The first woman admitted into Moscow State University, this Russian mathematician is known for her work on the theory of trigonometrical series. She wrote the books *Higher Algebra* and *The Theory of Series,* and her work is considered to be the standard for those who study functions and trigonometric series.

Bernoulli, Jakob (James, Jacques) (1654–1705) The oldest of the mathematical Bernoulli brothers, this Swiss mathematician is most renowned for his calculus, for SOLVING THE EQUATION $y^1 + P(x)y = Q(x)y^n$, now called the Bernoulli equation, and for an unfinished work that was finally published eight years after his death, entitled *Ars Conjectandi,* which, among other things, discusses the theory of PROBABILITY, and gives us the Bernoulli numbers but with little evidence of how he arrived at the series.

Bernoulli, Johann (John, Jean) (1667–1748) Swiss mathematician Johann Bernoulli, the younger brother of JAKOB BERNOULLI and a bitterly competitive man, proposed many mathematical problems to see if his brother, among others, was smart enough to solve them. One such challenge, the brachistochrone problem, led to Johann's fame as the founder of calculus of variations. Credit for his work in calculus was stolen from him by a private student, Guillaume de l'Hôpital, who published Johann's teachings in the form of the first calculus textbook ever written, entitled *Analyse des infiniment petits pour* *l'intelligence des lignes courbes.* But Johann also tried to steal credit for work on kinetic energy from another Bernoulli, his son Daniel, by putting a false publication date on his book *Hydraulica,* so that it would look like it was completed before Daniel's work, *Hydrodynamica,* on the same topic. Johann's arguments against SIR ISAAC NEWTON's theory of gravitation are considered to be singularly responsible for delaying the acceptance of Newton's work in the scientific world. Despite his personality flaws, acknowledgement for his mathematical mind has caused him to be called the Archimedes of his time.

Bhaskara (Bhaskara II, Bhaskaracharya) (1114–85) This 12th-century Indian mathematician was the author of six important works on math and astronomy, including *Lilavati,* on mathematics, and *Bijaganita* on algebra. His work included solving problems with NEGATIVE NUMBERS, rules for addition, subtraction, and multiplication, in which he showed two different ways of multiplying (both of which work, and neither of which are taught today), and rules for squaring and INVERSE OPERATIONS. He was the first person to use symbols to represent unknown numbers. He also came up with the first successful SOLUTION for the problem $nx^2 + 1 = y^2$, which later became known as Pell's EQUATION.

Bombelli, Rafaello (1526–72) This Italian engineer and architect was an amateur mathematician with no formal university education, yet he created a book on the subject of algebra, entitled *L'algebra.* He was the first to use symbols to denote EXPONENTS, and he was the creator of the concept of the IMAGINARY NUMBER, which he indicated as *i,* as the SOLUTION for the SQUARE ROOT of −1.

Boole, George (1815–64) As a young man in England, George Boole became fluent in Latin by the age of 12, in French, German, and Italian shortly thereafter, became an assistant elementary school teacher at the age of 16, and opened his own school at the age of 20. He ultimately wrote more than 50 papers and became the chair of mathematics at Queen's College in Cork, England. He was the first person to work on such algebraic number properties as the DISTRIBUTIVE PROPERTY, and he worked on differential EQUATIONS and probabilities. He is best known for his creation of the algebra of logic, now called

George Boole (Courtesy The Open Court Publishing Co., AIP Emilio Segrè Visual Archives)

BOOLEAN ALGEBRA, in which he used letters to represent statements that may be true or false, and which is the foundation of the math necessary for the creation of telephone switches and computers.

Brahmagupta (Bhillamalacarya) (ca. 598–670) India's most revered mathematician, Brahmagupta had many firsts in the study of algebra. He invented ZERO. He was the first to use algebra in astronomy, and he calculated solar and lunar eclipses and planetary motion. Brahmagupta created many mathematical rules for such ideas as negative and POSITIVE NUMBERS, multiplying using a place-value system, defining SQUARE ROOTS, the sum of SQUARES, the sum of cubes, and solving indeterminate QUADRATIC EQUATIONS. He was also the first to give the formula for the AREA of a cyclic QUADRILATERAL, now called Brahmagupta's THEOREM.

Briggs, Henry (1561–1630) British mathematician responsible for improving JOHN NAPIER'S LOGARITHMS, credited with assigning the VALUE of log 1 = 0, although he claims it was Napier's idea, and responsible for publishing a table of logarithms, *Arithmetica Logarithmica,* for the NATURAL NUMBERS 1 to 20,000 and 90,000 to 100,000. This table was completed by Adriaan Vlacq and published after his death as *Trigonometria Britannica.*

Brouncker, Viscount William (ca. 1620–84) Irish mathematician who entered Oxford University at the age of 16, earned a doctorate in medicine nine years later, became the first president of the Royal Society of London, and made contributions to mathematics through his work on LOGARITHMs, the quadrature of the HYPERBOLA, and CONTINUED FRACTIONS, as well as musical theory using algebra and logarithms. Through a case of mistaken identity, Euler named an EQUATION created by Brouncker for the wrong person, and it is instead called Pell's equation.

Cardan, Girolamo (Jerome Cardan, Girolamo Cardano) (1501–76) A colorful Italian physician, astronomer, inventor, writer, and mathematician, Cardan grew up assisting his father who tutored Leonardo da Vinci in geometry. Cardano wrote more than 200 pieces on a variety of topics, his most important work being his contributions to algebra. He was the first to publish on the topic of PROBABILITY, which he knew firsthand from many years of gambling away his money, his wife's jewelry, and the family's furniture. His best-known work, *Artis magnae sive de regulis algebraicis liber unus* (also known as *Ars Magna* or *The Great Art*), gives a SOLUTION to CUBIC and QUARTIC EQUATIONS, despite the fact that he learned about cubic equations from NICHOLAS TARTAGLIA and promised to never tell the secret of how he solved them. But upon discovering that SCIPIONE DEL FERRO, rather than Tartaglia, had been the first person to solve these, he published the results with a clear conscience. The cardan-shaft, a mechanical part that absorbs vertical movement in rear-wheel-drive cars, was invented by Cardan and first used in a carriage in 1548.

Cartwright, Dame Mary Lucy (1900–98) Devoted to mathematics and history, this British woman chose math because she felt it involved less work than memorizing facts and dates from history. Her work with John Littlewood solved the mystery behind problems with ratio and radar, and was the foundation for chaos theory.

Cataldi, Pietro Antonio (1548–1626) Italian who became a math teacher at the age of 17, and went on to write more than 30 math books on topics such as PERFECT NUMBERS, CONTINUED FRACTIONS, and algebra. His book, *Operetta di ordinanze quadre,* published in 1618, was specifically on the military applications of algebra.

Cauchy, Augustin-Louis (1789–1857) This famous French mathematician grew up with PIERRE-SIMON LAPLACE and JOSEPH-LOUIS LAGRANGE as family friends, and at the age of 13 began to study languages based on Lagrange's recommendation as a foundation for math. Cauchy contributed more work to the field of mathematics than any other person in history, writing 789 papers, many of which were more than 300 pages long. Some of his most famous topics include convergent and divergent infinite series, functions, differential calculus, astronomy, wave theory, the dispersion of light, and algebraic analysis.

Cavalieri, Bonaventura (1598–1647) The Italian mathematician famous for inventing the principle of indivisibles, published in 1635 in *Geometria indivisibilis continuorum nova,* and based

Augustin-Louis Cauchy
(Courtesy of AIP Emilio Segrè Visual Archives, E. Scott Barr Collection)

on work by ARCHIMEDES and Kepler, in which he initially proposed that a line is made up of an infinite number of points, and a PLANE is made up of an infinite number of lines, and that any magnitude, such as a line, or plane, or VOLUME, can be divided into an infinite number of smaller quantities. This idea was the beginning of infinitesimal calculus.

Cayler, Arthur (1821–95) British lawyer who was encouraged to study math as a child and eventually left law to became a professor of mathematics at Cambridge. He wrote and published more than 900 papers on mathematics, and is famous for developing work in algebra on matrices, as well as his work in geometry, which is now used in studying the space-time continuum.

Celsius, Anders (1701–44) Swedish inventor of the metric scale for measuring temperature, called the Celsius or centigrade scale, in which water freezes at 0° and boils at 100°.

Chrystal, George (1851–1911) Scottish mathematics professor who was very unhappy with the way algebra was being taught, so he wrote a book, entitled *Algebra,* that made him famous. It was praised for being the best book ever written on the subject.

Connes, Alain (1947–   ) French mathematician noted for his work in operator algebras and geometry, which led to opening up new areas of research in mathematics.

Cotes, Roger (1682–1716) British mathematician who published only one work in his short lifetime, dealing with a LOGARITHMic curve, is most famous for his THEOREM in trigonometry, and for editing SIR ISAAC NEWTON's second edition of *Principia.* Despite a continually deteriorating relationship between the two during the editing process, Newton believed "if he had lived, we might have known something."

Cramer, Gabriel (1704–52) Swiss mathematician who earned a doctoral degree at the age of 18 based on his thesis regarding the theory of sound, and became co-chair of mathematics at the Geneva university Académie de Clavin at the age of 20, in which he introduced the concept of teaching mathematics in French instead of Latin so that more people would have access to it. He is most famous for his published work on algebraic curves in 1750, entitled *Introduction á l'analyse des lignes* *courbes algébriques.* He wrote many papers and articles, including one that discussed the importance of multiple witnesses in legal cases based on PROBABILITY. He was so highly regarded by the leading mathematicians of his time that JOHANN BERNOULLI requested that after his death only Cramer should be allowed to publish his and his brother Jakob's work.

Dantzig, George (1914–) This American mathematician's early studies in mathematics ring of a true-life Hollywood movie. As a student arriving late to class one day, he copied two problems from the board and later turned in the "homework," only to find out that they were famous unsolved problems, for which he had worked out the SOLUTIONS. He ultimately became a professor emeritus at Stanford University, and is renowned as the "father of linear programming" for developing the simplex method of optimization, which is a process used for planning the use of resources, scheduling workers, and production planning.

Dedekind, Richard (1831–1916) German mathematician whose contributions to the field of mathematics include defining finite and INFINITE SETS, RATIONAL NUMBERS, and IRRATIONAL NUMBERS, and compiling editions of some of the most famous mathematicians of his time, including KARL FRIEDRICH GAUSS, and JOHANN PETER GUSTAV LEJEUNE DIRICHLET who was a close friend. Dedekind is most known for his work in number theory, specifically for his invention of the concept known as Dedekind cuts, in which he states that "when there is a cut (A_1, A_2) which is not produced by any rational number, then we create a new, irrational number a, which we regard as completely defined by this cut; we will say that this number a corresponds to this cut, or that it produces this cut." This idea was that every REAL NUMBER could divide rational numbers into two SUBSETs of numbers greater than and less than the real number, and to define irrational numbers in terms of rational numbers.

Descartes, René (1596–1650) French philosopher and mathematician who is known both as the founder of the modern school of mathematics and the founder of modern philosophy. Descartes's first major writing was *Le Monde,* which took four years to write, and explained his theory of

René Descartes (Engraving by W. Holl, from the original picture by Francis Hals in the gallery of the Louvre, Courtesy of AIP Emilio Segrè Visual Archives)

the structure of the universe. It was ready for publication in 1633, but news of GALILEO GALILEI's arrest was enough to keep Descartes from printing the book, and it remained unpublished until 14 years after his death. His major contributions to mathematics came in his next work, which included three books entitled *La Dioptrique, Les Météores,* and *La Géométrie* in which he is credited for inventing analytical geometry. He is responsible for the use of symbols to represent quantities; symbols for EXPONENTS; moving the TERMS of an EQUATION to one side of the equation; showing how to find indeterminate COEFFICIENTS to solve equations; and defining the elements for a point on a PLANE, using two fixed lines drawn at RIGHT ANGLES on a plane, each of which represent negative and positive VALUES, to determine the coordinates x and y of a point for an equation of a curve (he dealt only with curves), which are elements of the Cartesian GRAPH. Descartes claimed that he dreamt of analytical geometry one night in November 1619, and this was when he decided to pursue his ideas. He was known as having a cold disposition, and made it a personal mission to discredit PIERRE DE FERMAT. Descartes, who had been very sick as a young child, once told BLAISE PASCAL that the secret to maintaining his health was to sleep until 11:00 A.M. and to never let anyone make him get up before that hour. In 1650 Descartes moved to Stockholm, and met with the queen, at her request, every morning at 5:00. The walk to the palace in the cold air took its toll, and he died a few months later of pneumonia.

Diophantus of Alexandria (ca. 200–84) The Greek mathematician often called the "father of algebra," who was the first to use symbols, or VARIABLES, to represent unknown quantities. He wrote a 13-volume work entitled *Arithmetica,* which dealt with algebraic EQUATIONS and number theory, and contained 130 problems and their SOLUTIONS for determinate equations, as well as indeterminate equations, which are also known as Diophantine equations.

Dirichlet, Johann Peter Gustav Lejeune (1805–59) Belgian mathematician who at the age of 20 was the first to prove $n = 5$ when the number divisible by 5 is even, for Fermat's Last

Theorem. He also gave us the definition of a FUNCTION, which we still use today. Other important work includes analytic number theory, algebraic number theory, potential theory, boundary conditions in mechanics (now called the Dirichlet problem), and the Fourier series theory in trigonometry.

Euclid of Alexandria (ca. 325–265 B.C.E.) Egyptian mathematician, teacher, and writer about whom little is known. Some theories suggest that Euclid was the head of a team of mathematical writers, or that there was a team of such writers who called themselves Euclid without an actual person named Euclid involved. Most people, however, believe that Euclid was one man who is generally considered to be the greatest mathematician of the ancient world. He is famous for writing a series of 13 books called *The Elements,* which contain 465 propositions, or POSTULATEs, dealing with geometric algebra, SOLID geometry, PLANE geometry, number theory, the theory of proportions, and IRRATIONAL NUMBERs theory. His postulates include the ability to draw a straight line between two points, that all right angles are equal, and that only one line can be drawn through a point that is parallel to another given line. Euclid also included AXIOMs in his books, such as the idea that if things are equal to the same thing, then they are equal to each other. *The Elements* was used as a textbook to teach geometry for 2,000 years. It is said that Euclid's *Elements,* which has been printed in more than 1,000 editions, is the most translated and published book next to the Bible. Euclid also wrote other books, many of which have been lost, but some of which have survived through translations, including *Data* (properties of figures), *On Divisions, Optics, Phaenomena* (astronomy), and *Elements of Music,* which is attributed to Euclid, but not with certainty.

Eudoxus of Cnidus (ca. 408–355 B.C.E.) Greek scholar, mathematician, and astronomer who used geometry to explain the movement of planets. His major contributions to mathematics included the idea of REAL NUMBERs, the theory of proportion known in Euclid's *Elements* as the AXIOM of Eudoxus, which compares the lengths of rational or irrational lines, and his work on the method of exhaustion used later by ARCHIMEDES.

Leonhard Euler (Max-Planck-Institute of Physics, Courtesy of AIP Emilio Segrè Visual Archives)

Euler, Leonhard (Leonard Euler) (1707–83) Swiss mathematician noted for contributions in nearly every branch of mathematics. Euler is considered the greatest mathematician of the 18th century. Euler's father, who once roomed with JOHANN BERNOULLI at JACOB BERNOULLI's house, taught his son mathematics as a child. At the age of 14, Euler entered his father's alma mater, the University of Basel, to follow in his footsteps and study theology. While there, he convinced Johann Bernoulli to mentor him in mathematics, and within three years received a master's degree in philosophy, then gave up theology to study mathematics. He became a prolific writer on all major mathematical topics, publishing approximately 800 works. *Analysis Infinitorum,* published in 1748, covers trigonometry as its own branch of study, analytical geometry, and algebra, which became a basis for most algebra found in textbooks published since. In 1755, he wrote *Institutiones Calculi Differentialis,* considered to be the first "complete" textbook on differential calculus. His massive body of work includes algebra, differential calculus, calculus of variations, differential geometry, topology, trigonometry, ballistics, astronomy, shipbuilding, acoustics, elasticity, continuum mechanics, analytical mechanics, fluid mechanics, celestial mechanics, number theory, lunar theory, music theory, cartography, and geology. He was responsible for the introduction of beta functions, gamma functions, sin and cos as functions, and the notations $f(x)$, e, $i = \sqrt{1}$, Σ, and the universal acceptance of the symbol π for PI. Half of Euler's work was written after he became totally blind at the age of 59 in 1771. He amassed such a vast quantity of work that his unpublished papers continued to be printed for 50 years after his death.

Faltings, Gerd (1954–   ) German mathematician who was awarded the Fields Medal for his work in algebraic geometry, which was on a PROOF called the Mordell CONJECTURE. He also contributed to the final proof of Fermat's Last Theorem.

Fahrenheit, Gabriel Daniel (1686–1736) German physicist famous for inventing the mercury thermometer and for devising a temperature scale with UNITs measured in degrees Fahrenheit to determine when a liquid freezes and boils.

Fefferman, Charles Louis (1949–) American mathematician who began college at the age of 12, graduated at 17, received his doctorate at 20, and at 22 became a professor at the University of Chicago, making him the youngest professor in American history. Currently a professor at Princeton University, he is famous for his Fields Medal award-winning work on partial differential EQUATIONS.

Fermat, Pierre de (1601–65) French lawyer and government official who studied mathematics as a pastime and became one of the most highly regarded mathematicians of his time, although a bit controversial. Fermat never published his work, but instead wrote letters to prominent mathematicians challenging them with problems that he had already secretly solved. This often resulted in accusations of purposely taunting them with "impossible" problems. He and RENÉ DESCARTES were at odds for many years because of Fermat's low opinion of Descartes's work, which he once described as "groping about in the shadows," and because Descartes was on a personal mission to ruin Fermat's reputation as a mathematician. Fermat is most known for his work in number theory, in which he states that $x^n + y^n = z^n$ has no nonzero INTEGER SOLUTIONS for x, y, and z when $n > 2$, for which he gave no PROOF. This is known as Fermat's Last Theorem, and at the time was not considered important enough by mathematicians to try to find a proof. The THEOREM remained unsolved for more than 300 years, and was finally proved by British mathematician ANDREW JOHN WILES in 1994.

Pierre de Fermat (University of Rochester, Courtesy of AIP Emilio Segrè Visual Archives)

Ferrari, Lodovico (Ludovico Ferraro) (1522–65) This Italian mathematician began his mathematical studies as a 14-year-old servant to GIROLAMO CARDAN, who decided the boy was so bright that he had to teach him. Ferrari discovered the SOLUTION to the QUARTIC EQUATION, but because he was working with Cardan and the solution involved the CUBIC equation, he ended up in a very public and lengthy series of insults with NICHOLAS TARTAGLIA. The two ultimately had a public debate, which Ferrari clearly was winning, so Tartaglia left town.

Ferro, Scipione del (Ferreo, dal Ferro) (1465–1526) Italian mathematician best known for being the first to solve CUBIC

EQUATIONS, in which he used a formula that was much like the one used by Babylonians to solve QUADRATIC EQUATIONS. GIROLAMO CARDAN called del Ferro's equation "a gift from heaven."

Fibonacci, Leonardo Pisano (Leonardo of Pisa, Leonardo Bigollo, Leonardo Bonacci, Leonardo Bonaccio, Leonardo Bonacij) (1170–1250) As a boy, this Italian traveled extensively with his father and learned, through his travels, about "the art of the Indians' nine symbols," which was the Hindu-Arabic number system we use today. Fibonacci is most famous for his first book *Liber Abbaci* (*Book of the Abacus* or *Book of Accounting*) printed by hand and published in 1202 (printing presses were not yet invented). Written in Latin, this book explained addition, subtraction, MULTIPLICATION, and division, using problems as examples, in the Hindu-Arabic number system. It was responsible for bringing this number system with the symbol ZERO (he called it zephirum) to Europe as a replacement for the Roman numeral system in use at the time, which has no symbol for zero. It also introduced the famous FIBONACCI SEQUENCE, in which a given number is the SUM of the two numbers immediately before it. Fibonacci wrote several other books, including *Practica Geometria,* geometry problems based on EUCLID OF ALEXANDRIA's work, *Flos,* dealing with the approximation of ROOTs, and *Liber Quadratorum,* on number theory, all of which have handwritten copies still in existence.

Flügge-Lotz, Irmgard (1903–74) German mathematician and engineer who studied differential EQUATIONs and developed the Lotz method for determining the lifting force on an aircraft wing. In 1961, she became the first woman professor of engineering at Stanford University.

Fourier, Jean-Baptiste-Joseph (1768–1830) French mathematician who left his study of the priesthood for algebra, was nearly beheaded for speaking out during the French Revolution, and soon thereafter came under the service of Napoleon as scientific adviser and founder of the Cairo Institute. Fourier is known for his work on linear inequalities, and for Fourier's THEOREM, regarding the roots of an algebraic EQUATION. While working under Napoleon, Fourier developed his theory

Jean-Baptiste-Joseph Fourier (Courtesy of AIP Emilio Segrè Visual Archives)

of heat, and wrote a paper entitled *On the Propagation of Heat in Solid Bodies* in 1807, which he later expanded to include cooling of infinite solids, terrestrial heat, and radiant heat. This theory was controversial through his lifetime, yet ultimately proved important in trigonometry and the theory of functions of a real VARIABLE.

Frobenius, Ferdinand Georg (1849–1917) German mathematician who made many contributions to the field. He is best known for his work on abstract algebra and group theory, which is used today in quantum mechanics, and he is regarded as the one who started the study of representation theory of groups. He is highly regarded for his calculations in his work, particularly singled out for his "algebraic approach."

Galilei, Galileo (1564–1642) The Italian scientist most known for his controversial views relating to astronomy, Galileo was also a professional musician, a mathematician, physicist, student of medicine, and philosopher. Galileo observed and articulated the idea of inertia, upon which SIR ISAAC NEWTON's first law of motion is based, and is credited as the first scientist to use experimentation to back up his conclusions. He was mathematics professor at Pisa, where he challenged Aristotle's ideas on the speed at which an object falls based on its weight. After his professorship was not renewed, he became chair of mathematics at the University of Padua from 1592 to 1610, where he worked on motion and parabolic paths of projectiles, and developed a telescope by mathematical computations based on the stories he had heard of a telescope invented in Holland that same year. For this impressive invention, he was named court mathematician and philosopher in Florence. This work with the telescope was the beginning of his demise, as he soon observed that the Earth could not possibly be the center of the universe, despite Aristotle's teachings. He published his theories in a book called *The Starry Messenger,* which he wrote in Italian rather than Latin so that more people would have access to it. The Roman Catholic Church banned the book and ordered him to stop teaching his views, but he published another book, *Dialog Concerning the Two Chief World Systems,* in 1632, attacking the beliefs of both Aristotle and PTOLEMY. The Inquisition accused him of heresy, and at

Galileo Galilei (Courtesy of AIP Emilio Segrè Visual Archives)

the age of 70 he was forced to recant his beliefs under threat of death by torture. He was condemned to house arrest, and was not allowed to publish any work. He secretly continued his research and wrote anyway, and his work was smuggled into the Netherlands and published in 1638. Galileo died, blind, under house arrest, on Christmas Day in 1642, the same day Newton was born.

Galois, Evariste (1811–32) French mathematician known for his "genius and stupidity." He published his first mathematical paper at the age of 17 on CONTINUED FRACTIONS. By the age of 18, Galois was imprisoned for six months for wearing a National Guard uniform, defying French law. At 19, he wrote three important papers on algebraic EQUATIONS. He is most famous for the legendary story of the last 24 hours of his life, despite the fact that it is generally acknowledged that the story has been somewhat exaggerated. According to legend, the 20-year-old Galois was challenged to a duel, which it is thought was either instigated to defend his politics, to defend the honor of a woman, or as part of an elaborate conspiracy to get rid of him for his radical views. Afraid that he might lose the duel, he wrote 60 pages of mathematical concepts that he wanted to get on paper, including group theory, scribbling frequently and frantically in the margins that he didn't have enough time as he worked through the night. He was right. The next day he was shot in the abdomen, and he died the day after. The importance of the theory he frantically wrote down was that he had discovered "under what conditions an equation can be solved," which is now known in geometry as the Galois theory.

Gauss, Karl Friedrich (1777–1855) German mathematician who is often referred to as the "prince of mathematics." He is famous for several accomplishments made by the time he reached his early 20s, including correcting one of his father's payroll calculations at the age of three, instantly SUMMing an ARITHMETIC SERIES of the INTEGERS of 1 to 100 in the third grade (the answer is 5,050), constructing a heptadecagon for the first time in history using just a straightedge and a compass at the age of 19, proving the FUNDAMENTAL THEOREM OF ALGEBRA at the age of 20, and discovering the Fundamental Theorem of Arithmetic at the age of 24, at which time he also

Karl Friedrich Gauss
(Courtesy of AIP Emilio Segrè
Visual Archives)

published his book *Disquisitiones Arithmeticae,* which was a breakthrough in number theory that explained the use of imaginary and congruent numbers. He published little, because of his desire to constantly improve the work he had already done, but went on to make many more contributions in mathematics, particularly in geometry, applied mathematics, astronomy, in which he calculated the correct orbit for the new asteroid Ceres, physics, and statistics. He also coinvented the first working telegraph.

Germain, Marie-Sophie (1776–1831) French mathematician and physicist who is known as the HYPATIA of the 18th century, she became interested in mathematics when she read about ARCHIMEDES' death in *History of Mathematics,* a book by Jean-Étienne Montucla that she found in her father's library. Because she was a girl, she was not allowed to study mathematics as a child, and her parents took away her candles and heat at bedtime to keep her from reading Newton and Euler. It did not work, and she instead stayed up and taught herself calculus. She accomplished her most famous contributions to the sciences by assuming the identity of a male student, Monsieur Le Blanc, who had left Paris, and studying mathematics at the Ecole Polytechnique with JOSEPH-LOUIS LAGRANGE by remotely sending letters and turning in papers using LeBlanc's name. She is remembered for her work on FERMAT'S LAST THEOREM, in which she devised a unique way of looking at the problem, and her theory of elasticity, which deals with stress on construction materials such as steel beams.

Girard, Albert (1592–1632) French musician, engineer, and mathematician who is credited as the first to define the FIBONACCI SEQUENCE as $f_n+2 = f_n+1 + f_n$. He also made contributions to the FUNDAMENTAL THEOREM OF ALGEBRA, which was not defined enough to prevent subsequent mathematicians from attempting a PROOF that could not work, and which was finally proved 174 years later by KARL FRIEDRICH GAUSS.

Goldbach, Christian (1690–1764) Russian historian and mathematician who studied curves, the theory of EQUATIONS, number theory, and infinite SUMs. He is most known for his letter to LEONHARD EULER, in which he wrote that every EVEN NUMBER greater than two is the sum of two primes, although he gave no PROOF. Known as Goldbach's CONJECTURE, this remains one of the most long-standing unsolved mathematical problems of all time.

Gregory, James (1638–75) This Scottish mathematician, who was taught geometry by his mother, used mathematical computations to invent the first reflecting telescope, now called the Gregorian telescope. His book, *Optica Promota,* contained POSTULATEs, definitions, and 59 THEOREMs pertaining to the reflection and refraction of light, and explained how the telescope worked in detail, although he had no way to build it. He also made contributions in geometry, particularly regarding the AREA of a CIRCLE and a HYPERBOLA, and in trigonometry. The majority of his work went unpublished because of a dispute in which he was accused of stealing another mathematician's work, but he is credited as having made many of the same major findings in mathematics as SIR ISAAC NEWTON, Brook Taylor, Johannes Kepler, AUGUSTIN-LOUIS CAUCHY, and BERNHARD RIEMANN prior to their own discoveries.

Harriot, Thomas (1560–1621) British explorer, astronomer, and mathematician who never published any of his work during his lifetime. Harriot began his career working for Sir Walter Raleigh's expeditions to the New World. Harriot worked on optics, studied curves and motion, was interested in the theory for the rainbow, observed and recorded Halley's comet leading to the computation of its orbit, made the first observations with a telescope in England, made the first telescopic drawing of the Moon, and was the first to discover sunspots and calculate the Sun's rotation. He proposed a problem and CONJECTURE to Johannes Kepler, which he derived from a similar problem asked of him by Raleigh regarding the densest stacking of cannonballs, which was not proved until 1998. His contributions to algebra include SIMPLIFYing algebraic notation, COMPLETING THE SQUARE, and the invention of symbols for "greater than" and "less than," which is sometimes credited to his book editor.

Hérigone, Pierre (Herigonus) (1580–1643) French mathematician about whom little is known except for his published work on elementary mathematics called *Cursus Mathematicus,* which he wrote in Latin and French, and which introduced a complete SET of mathematical symbols. His symbols for EXPONENTs, a, a2, a3, and so on, were the predecessors to the ones we use today, a, a^2, a^3, and so on, which were subsequently introduced by RENÉ DESCARTES.

Hermite, Charles (1822–1901) French mathematician known for having "a kind of positive hatred of geometry," according to his student, mathematician Jacques Hadamard, was most famous for his passion for analysis, which led him to discover that while an algebraic EQUATION of the fifth degree could not be solved in radicals, it could be solved using elliptic functions. Hermite POLYNOMIALs and Hermite's differential equation are among his namesakes for his many contributions to the science of mathematics, which include elliptic functions, number theory, and polynomials.

Hipparchus (Hipparchus of Nicaea, of Rhodes, of Bithynia) (ca. 190–20 B.C.E.) Greek astronomer and mathematician known as the "lover of truth" for the accuracy of his calculations, and as the inventor or father of trigonometry.

He was the first to describe a location on the planet according to its longitude and latitude, and the first to introduce that a CIRCLE is divisible into 360°. He calculated the solar year to measure 365 $1/4$ days, minus 4 minutes, 48 seconds, which is off by just six minutes, and the lunar year to be 29 days, 12 hours, 44 minutes, 2 $1/2$ seconds, which is off by less than one second. His only surviving work is a three-volume book entitled *Commentary on Aratus and Eudoxus.* PTOLEMY's work is believed to be based largely on Hipparchus's, and many believe that the Ptolemaic Theory should really be named the Hipparchian Theory.

Holywood, John de (Johannes De Sacrobosco, John of Hollywood, John of Halifax) (ca. 1200–56) This medieval English mathematician and astronomer wrote several books that became texts, including *Tractatus algorismi* on arithmetic, *Tractatus de Sphera* on astronomy, which was used for four centuries, *Tractatus de Quadrante* on quadrants, and *De Anni Ratione* on time and the calendar.

Hypatia of Alexandria (ca. 370–415) Egyptian astronomer, philosopher, teacher, and mathematician regarded as the first woman scientist, and the first woman to contribute to the study of mathematics. As head of the Platonist school in Alexandria, in addition to teaching through lectures, she wrote about the work of others, including DIOPHANTUS's *Arithmetica,* Apollonius and PTOLEMY's work, and with her father on EUCLID's *Elements.* Her mathematical abilities were exceptionally helpful in her work on astronomy, particularly in her construction of an astrolabe to measure the positions of heavenly bodies. Since science was, at her time, related to paganism by the Christians then coming into power, Hypatia was declared an evil enemy of the church, and was subsequently attacked and murdered by a mob, which many historians believe marked the beginning of the end of Alexandria's reign as a major center of scientific learning.

Ibn Ezra, Abraham ben Meir (Rabbi Ben Ezra, Aben Ezra) (1092–1167) This Jewish scholar lived in Spain and traveled extensively. He was known as a great poet, but his scientific interests led to writing about mathematics, particularly *The Book of the Unit* and the *Book of the Number.* Both writings dealt with Indian mathematics, specifically the numbers one through nine, the concept of ZERO, and the DECIMAL system. His work made it to Europe, yet it was several centuries before it became accepted practice.

Khayyam, Omar (Ghiyath al-Din Abu'l-Fath Umar ibn Ibrahim Al-Nisaburi al-Khayyami, Omar Alkayami) (1048–1131) Iranian mathematician, astronomer, and poet who wrote books on music, algebra, and arithmetic before the age of 25, and who is known as the first to develop a general theory of CUBIC EQUATIONS. He also worked with RATIOS, and accidentally proved non-Euclidean geometry properties of figures. His poetry is found in the famous *Rubaiyat,* and he is often better known for his poetic writings than for his work in mathematics.

al-Khwārizmī, Abu Ja'far Muhammad ibn Musa (ca. 780–850) The word *algebra* was derived from this Arabian mathematician's famous writing, entitled *Al-gebr we'l mukabala* (or *Hisab al-jabr w'al-muqabala*), considered the most important work in the history of mathematics. *Al-gebr* means restoration, and refers to adding and subtracting from both sides of the EQUATION, and *al mukabala* means simplification, which deals with COMBINING LIKE TERMS. Al-Khwārizmī explained QUADRATIC EQUATIONS, and gave rules for addition and subtraction for unknown quantities, SQUARES, and SQUARE ROOTS, among others. His work was the first to introduce the DECIMAL system and the concept of ZERO. Its translation in the early 12th century into Latin brought these concepts to the rest of the world, and helped define the study of algebra as we know it today.

Kneser, Hellmuth (1898–1973) Russian mathematician credited with making contributions in many areas of mathematics. He refused to specialize in just one area, and his works include differential geometry, non-Euclidean geometry, sums of SQUARES, analytic functions, economic theory, quantum theory, and game theory, among others.

Koopmans, Tjalling Charles (1910–86) Dutch-American physicist, mathematician, and economist whose important contributions were in applying mathematics to real-world applications. He was corecipient of the Nobel Prize, with Soviet mathematician Leonid

Kantrovich, for independently developing the work of linear programming, a method of mathematically solving the allocation of resources. He did this work to figure out shipping schedules for the British Merchant Shipping Mission during World War II. As a physicist, he developed Koopmans' Theorem in quantum mechanics before turning to economics, supposedly because pure mathematics was not interesting or rewarding enough.

Kovalevskaya, Sofia Vasilyevna (Sofya Krovin-Krukoskaya, Sonja Krovin-Krukoskaya) (1850–91) Russian mathematician who overcame prejudice against women to make substantial contributions to the subject of mathematics. She submitted a paper entitled "On the Rotation of a Solid Body about a Fixed Point" to the French Academy of Science for the Prix Bordin competition, and it became the first SOLUTION for a body that was asymmetrical and in which the center was not on an AXIS. This work resulted in winning both the competition and an increase in prize money for its brilliance. Kovalevskaya's education started young. At age 11, she learned mathematics from her bedroom walls, where lecture notes on calculus had been pasted up as wallpaper, and she taught herself trigonometry at the age of 14 to understand a physics book given to her family as a gift from the author, their neighbor. Despite her father's attempts to prevent her from studying mathematics, because it was interfering with her other studies and because she was a girl, her neighbor convinced her father that she was gifted and should be allowed to learn. After much struggle with educational leaders because of her gender, she eventually earned a doctorate in mathematics despite being denied permission to attend the university, thanks to the help of KARL THEODOR WILHELM WEIERSTRASS, who recognized her brilliance and tutored her privately. She was not allowed to teach at the university level, however, and was offered only a position of teaching arithmetic to elementary school girls. She ultimately became a professor in Stockholm three years after winning the Prix Bordin.

Lagrange, Joseph-Louis (Giuseppe Lodovico Lagrangia) (1736–1813) Italian mathematician who was mostly self-taught. Lagrange is considered one of the premier mathematicians of his century. His work was for the most part in calculus, and his brilliance was recognized early in his

Joseph-Louis Lagrange (Courtesy of AIP Emilio Segrè Visual Archives, E. Scott Barr Collection)

career by LEONHARD EULER, who helped him secure a position as a mathematics professor at the age of 19. Lagrange's work included calculus of variations, calculus of probabilities, fluid mechanics, PROBABILITY, astronomy, and number theory. He is known for using algebra in his work in mechanics and differential calculus. He wrote an AVERAGE of one mathematical paper a month over a 20-year period.

Laplace, Pierre-Simon (1794–1827) A French mathematician, astronomer, and physicist, Laplace is known for his original contributions to the sciences. Laplace's most famous published works involve mechanics, physics, geometry, and astronomy, and include *Exposition du systeme du monde,* a collection of five books regarding the history of astronomy and his theories on the planetary motion, gravity, and his nebular HYPOTHESIS; the *Triaté du mécanique céleste,* another five-volume set dealing with mechanics of planetary motion with mathematical EQUATIONS; and *Théorie analytique des probabilités,* two books on functions and PROBABILITY. His revised volume of *Mécanique céleste* included the idea of analyzing reactions between molecules, which made a great contribution to the field of physics.

Pierre-Simon Laplace
(Courtesy of AIP Emilio Segrè Visual Archives)

Leibniz, Gottfried Wilhelm (1646–1716) German scholar equally known for his literature, philosophy, and mathematics. Leibniz's controversial fame surrounds SIR ISAAC NEWTON and the invention of calculus. Leibniz started his college education at the University of Leipzig at the age of 14, received his bachelor's degree at the age of 16, and then began working toward a law degree, first earning his bachelor's and then his master's four years later. He was considered a Renaissance man despite the timing because of his in-depth interest and knowledge in so many areas of study, and he is known to have regularly corresponded with 600 people in his quest for knowledge. In 1676 he became librarian and court counsel for the Duke of Hanover, positions he held for the rest of his life. He was individually responsible for being the first to introduce the terms COORDINATES and AXES of coordinates, and in 1694 directly influenced the creation of the Berlin Academy. The Newton controversy involved the timing of their publications on calculus. In question was whether Leibniz had stolen his ideas about calculus from letters he and Newton exchanged. Many, including Newton, believed this to be the case. JOHANN BERNOULLI jumped to Leibniz's defense, making it a personal mission to discredit Newton's character, beginning with writing a slanderous letter and then shortly thereafter claiming that he actually never wrote it. Newton, known for his outrageous temper, was in a position of power as president of the Royal Society at the time of the accusation against Leibniz, and he took advantage of this by appointing a committee that would decide once and for all who really invented calculus. Based on the committee's unsurprising finding that it was Newton, he anonymously wrote the society's official decision, and followed that up with an anonymous review of its decision that Newton was the true inventor of calculus. The debate did not end there, as scholars argued the case well after the deaths of both men. Leibniz and Newton are now considered by most to have created their ideas independently, and Leibniz is individually credited with inventing differential calculus.

Lovelace, Augusta Ada (Augusta Ada Byron, Ada Byron Lovelace, Ada Byron King, Countess of Lovelace)

(1815–52) This British countess, daughter of the famous poet George Gordon, Lord Byron, was educated by private tutors and lived the privileged life of high society. She is most known for her friendship and work with mathematician CHARLES BABBAGE. Lovelace was a proponent of Babbage's Analytical Engine, and suggested how he might make the machine calculate the BERNOULLI NUMBERS. For this idea, many credit her for creating the first computer program.

Lucas, François-Edouard-Anatole

(1842–91) French mathematics professor who is most widely known for creating the formula for the FIBONACCI SEQUENCE, and for devising a companion formula now known as the Lucas SEQUENCE, or Lucas numbers, in which the first number is 1, the second 3, and the following numbers behave the same as the Fibonacci sequence. In 1876 he discovered the largest PRIME NUMBER that has ever been calculated without the use of a computer. Called a Mersenne number, it is denoted as $M_{127}(2^{127} - 1)$ and equals 170,141,183,460,469,231,731,687,303,715,884,105,727. Lucas invented a famous mathematical puzzle called the Tower of Hanoi, and wrote a classic book on "recreational" mathematics called *Récréations mathématiques.*

Maclaurin, Colin (1698–1746) Scottish mathematician who entered Scotland's University of Glasgow at the age of 11 and became devoted to mathematics after finding a copy of EUCLID's *Elements* in another student's room. Maclaurin received his M.A. degree from the university at the age of 14 with a thesis that expanded on Newton's ideas, entitled *On the Power of Gravity.* In 1719 he traveled to London and became a member of the Royal Society, and soon became friends with Newton. He worked to expand the Medical Society of Edinburgh to include other branches of education, which ultimately became the Royal Society of Edinburgh. His most important works are *Geometrica Organica,* published in 1720, his 763-page *Treatise of Fluxions* regarding Newton's calculus, and his *Treatise on Algebra,* published in 1748. Unlike many of the mathematicians of his time, he was highly regarded as a kind and caring person who made himself available to anyone interested in the ideas of mathematics.

Mahavira (Mahaviracharya) (ca. 800–70) Indian mathematician whose only known book, *Ganita Sera Samgraha,* is the first Indian text dedicated solely to the study of mathematics. Its nine chapters include work on arithmetical OPERATIONS, the place-value number system, FRACTIONS, squaring numbers, INDETERMINATE EQUATIONS, calculating the VOLUME of a SPHERE, CUBE ROOTS, and AREA. This book, an elaboration on BRAHMAGUPTA's work, is considered all-inclusive of the mathematics known during the ninth century.

Mandelbrot, Benoit (1924–   ) French-born American mathematician who spent his early years learning math mostly on his own due to World War II, and for which he credits his success in being able to think in unconventional ways. His greatest contribution is in geometry. He is credited with creating the field of fractal geometry, which he explained in his book *Les objects fractals, forn, hazard et dimension,* published in 1975, and in his 1982 publication *The Fractal Geometry of Nature.*

Mori, Shigefumi (1951–   ) Japanese mathematician dedicated to classification of algebraic varieties in algebraic geometry, which includes the study of curves as related to POLYNOMIALS. Mori won the prestigious Fields Medal in 1990 for work he did over a period of 12 years, finding SOLUTIONS that many felt were unattainable. Other awards include the Cole Prize in Algebra from the American Mathematical Society, also in 1990, the Japan Mathematical Society's Yanaga Prize, and the Chunichi Culture Prize.

Müller, Johann *See* REGIOMONTANUS.

Napier, John (Jhone Neper, Jhone Napare) (1550–1617) The Scot known for his religious convictions and his book *Plaine Discovery of the Whole Revelation of St. John,* Napier spent most of his time in politics and religious controversies, studying mathematics only as a hobby. He began his higher education at St. Andrews University at the age of 13. Napier is most famous as the inventor of LOGARITHMS, which he completed over many years in his spare time. He also introduced the idea of DECIMAL notation for FRACTIONS. In 1617 he invented a system for multiplying and dividing using numbering rods, which were ivory sticks marked with numbers that, when placed next to each other, showed the PRODUCT or QUOTIENT. Because the ivory sticks resembled bones, they were dubbed Napier's bones. He elaborated on the rods and invented two more for determining SQUARE and CUBE ROOTS. During his lifetime, because of his unique ways of thinking and his eccentric behavior, many thought he was a warlock in league with the devil.

Newton, Sir Isaac (1642–1727) A British mathematician and scientist, Newton survived an unhappy and uneducated childhood to become one of the most famous scientists of all time, and the bulk of his most brilliant work took place during an unlikely two-year period in his early twenties. Newton's illiterate and uneducated, but wealthy, father died before Newton was born, and the child was shuffled between relatives until the age of 10, when he started elementary school. Due to an obvious lack of any ability, he was pulled out of school shortly after he started, and was finally allowed to go back again at the age of 13, at which point he was sent to live with the school's headmaster. Newton enjoyed his studies so much this time around that he actually felt it was a sin. Unlike many of his contemporaries, for example GOTTFRIED WILHELM LEIBNIZ, Newton did not start university until the late age of 18, entering Trinity College, Cambridge, in 1661. Despite his mother's wealth, his status was that of a "sizar," one who was financially subsidized by the college, but who in exchange was expected to work as a servant for the more wealthy students. In 1663, Newton picked up EUCLID's *Elements* for the first time, and soon began to devour mathematics books by WILLIAM OUGHTRED, RENÉ DESCARTES, FRANÇOISE VIÈTE, JOHN WALLIS, and others. In 1665, after an unspectacular four years in college, Newton graduated, and the university suddenly closed because of the plague. Newton returned to his family's farm at the age of 22. Over the next year and a half, away from academia, Newton's mind was free to brew up the scientific ideas for which he became famous. By the age of 24, Newton had created calculus with his theory of fluxions, and had discovered the laws of motion and the universal law of gravitation. In 1667 Newton returned to the university (now reopened) to work on his master's degree, and reluctantly began to share his ideas. Then in 1669, as Lucasian Professor, he began his work in optics. Declaring that white light is actually made up of a spectrum of colors, and concerned that this could cause problems with refracting telescopes, he built a reflecting telescope. In 1672 he donated it to the Royal Society, which consequently elected him as a member and allowed him to publish his first paper on the properties of light. Controversy soon followed. Two mathematicians, Robert Hooke and Christiaan Huygens, disagreed with Newton's science, and Hooke went so far as to accuse Newton of stealing his research. The Jesuits also disagreed with his science, for different reasons, and he soon received a series of violent letters. These attacks to his character and his work were more than he could stand, and in 1678 Newton suffered a nervous breakdown. In 1684 Newton began a serious dialogue with Edmond Halley about his ideas in physics and astronomy, and Halley convinced him to publish. In 1687 Newton came out with *Philosophiae Naturalis Principia Mathematica,* or the *Mathematical Principles of Natural Philosophy.* Known as *Principia,* this work became Newton's masterpiece. The following year, politics and religion began to interfere with the educational mandates of the university, and King James II began to appoint unqualified Catholics to every opening within the school, which Newton strongly and publicly opposed as an attack on education. When William of Orange

Isaac Newton (Original engraving by unknown artist, Courtesy of AIP Emilio Segrè Visual Archives)

overthrew King James's regime, Newton was rewarded for his stand by being appointed to Parliament in 1689. Four years later, in 1693, his career in scientific research came to a crashing halt when he suffered another nervous breakdown. Within six years he held a new position as head of the Royal Mint, and he became a very wealthy man. In 1703 Newton was elected president of the Royal Society, and in 1704, after Hooke's death, he published his work *Optiks,* which became a significant contribution to the field. In 1705 he was knighted by Queen Anne for his contributions to science, the first scientist ever to receive such an honor. In 1711 controversy struck again: An article in the *Transactions of the Royal Society of London* accused Leibniz of plagiarizing Newton's work, claiming that Newton, not Leibniz, invented calculus. As president of the society, Newton established an "impartial committee" to investigate, and given the leadership, the results were not surprising. Newton anonymously wrote the society's official decree in favor of himself, and never gave Leibniz the chance to support his side of the story. The incident became an international affair, and was never settled during the lifetime of either man.

Nicomachus of Gerasa (Thrasyllus) (ca. 60–120) Greek mathematician whose book *Introduction to Arithmetic (Arithmetike Eisagoge),* was the first text to focus solely on arithmetic without geometry. It dealt mainly with number properties and theory, including EVEN and ODD NUMBERS, PRIME NUMBERS, PERFECT NUMBERS, FRACTIONS, ABUNDANT NUMBERS, DEFICIENT NUMBERS, and RATIOS. Translated into Latin in the year 175, it became the standard arithmetic textbook for the next 1,000 years, despite the fact that it contained errors, had no PROOFS, and merely stated its theorems with the addition of a few examples. Two other books include his two-volume *The Theology of Numbers (Theologoumena Arithmetikes),* which covered the "mystic properties" of numbers, and the *Manual of Harmonics,* devoted to proving musical theory through the use of mathematical THEOREMS.

Nightingale, Florence (1820–1910) A British reformist known for her pioneering work in hospital sanitation and nursing, Nightingale developed a system of statistical analysis that revolutionized the medical world. Nightingale began her mathematical studies at the age of 20, and eventually became a mathematics tutor. Although nursing was considered a disreputable career, she enrolled at the Institute of St. Vincent de Paul in Egypt in 1850 because she felt it was her calling from God to do so. In 1854 she went to work in the British military hospital in Constantinople, Turkey. The conditions she encountered prompted her to start collecting statistical data on the mortality rate of soldiers, and she calculated that improved sanitation would reduce the number of deaths. Nightingale's work resulted in a reduction of the mortality rate from 60 percent to just over 2 percent in a six-month period. She developed a graph of her data, called the Polar Area Diagram, and further used the information to compute that the British army would be completely wiped out by disease given the then-current rate of soldier deaths. Her mathematics gave her the ability to push for reform of hospital conditions, as well as institute training for nurses, which ultimately changed the reputation of the profession. For her work, she received many honors, including election as the first woman Fellow of the Royal Statistical Society and as an honorary member of the American Statistical Association.

Noether, Amalie (Emmy) (1882–1935) A German mathematician touted by Einstein as "the most significant creative mathematical genius thus far produced since the higher education of women began." He placed her amongst the most gifted mathematicians ever in the field of algebra. Her major contributions to the sciences include the development of group theory in physics, as well as her book *Idealtheorie in Ringbereichen,* which dealt with ring theory. She made discoveries in theoretical physics, among them Noether's Theorem, which was praised by Einstein and ultimately aided him in the development of the theory of relativity. Despite her genius, she spent most of her life working in mathematics unpaid because of her gender. With the onset of Nazi power in Germany in 1933, she moved to Bryn Mawr College in the U.S., and taught there until her death.

Oresme, Nicholas (Nichole d'Oresme, Nicholas Oresmus) (1323–82) French mathematician who invented COORDINATE geometry some 200 years before the birth of RENÉ DESCARTES, and proposed using a GRAPH for plotting VARIABLES. He taught against Aristotle's astronomy more than 100 years before the birth of Copernicus, but ultimately decided that his ideas about the motion of the Earth were wrong. He also worked on FRACTIONAL EXPONENTS and the science of light.

Oughtred, William (1574–1660) A British theologian who tutored students in mathematics, including JOHN WALLIS, Oughtred is famous for his book *Clavis Mathematicae,* which introduced the symbol x for MULTIPLICATION and covered Hindu-Arabic numbers and DECIMAL FRACTIONS. He is also known for inventing the predecessor to the slide rule, and is often credited as the inventor of the circular slide rule, which he wrote about in his book *Circles of Proportion and the Horizontal Instrument,* although there is some argument that another mathematician might have invented it prior to Oughtred.

Pacioli, Lucas (Luca Pacioli, Lucas de Burgo, Lucas Paciolus) (ca. 1445–1517) Italian theologian and mathematician whose importance to the field of mathematics lies in the books he compiled of the work of others. He wrote three books that were intended for his students and were unpublished, in 1470, 1477, and 1480, of which only one still exists. His first published work, *Summa de Arithmetica, Geometria, Proporzioni e Proporziolalita,* a 600-page book published in 1494, is considered his most famous work, and included everything known about mathematics at the time, with information on arithmetic, algebra, geometry, money, weights and measures, bookkeeping, and gaming. He published *Divina Proportione* in 1509, a series of three books dealing with the golden RATIO, also known as the divine proportion, some of Euclid's THEOREMS, and POLYGONS. This book was illustrated by his close friend Leonardo da Vinci.

Pascal, Blaise (1623–62) A French mathematician forbidden to study mathematics as a child, Pascal went on to make important contributions to the disciplines of algebra and geometry. Pascal saw his first mathematics book, Euclid's *Elements,* at the age of 12. At 14 he began to attend weekly meetings with some of France's visionary mathematicians, who eventually went on to start the French Academy. His first geometry paper, regarding conics, was written at the age of 16 and ultimately gave us a

THEOREM known as Pascal's Theorem when published in 1779. At 18 he invented the first digital calculator. Pascal was intrigued with experiments on atmospheric pressure and vacuums. He published *New Experiments Concerning Vacuums* in 1647, but this work was not well received by the scientific community. In fact, after a two-day visit with Pascal, RENÉ DESCARTES wrote a letter to a fellow mathematician saying that Pascal had "too much vacuum in his head." Pascal's law of pressure, explained in his *Treatise on the Equilibrium of Liquids,* published in 1653, is considered a major contribution to physical theory. In this same year, he came up with his idea for the arithmetical triangle, now known as PASCAL'S TRIANGLE, published in his *Treatise on the Arithmetical Triangle.* The following year, Pascal began writing to PIERRE DE FERMAT about an idea he had, and in the course of five letters between them, they came up with the theory of PROBABILITY. Pascal became deeply religious when his father died, and after nearly suffering a fatal accident himself at the age of 31, coupled with ill health and severe pain that plagued him most of his life, he abandoned mathematics and devoted himself to religion. His philosophical book on life and death, *Pensées,* written between 1656 and 1658, nonetheless showed his natural bias towards mathematics. In it, he stated that "if God does not exist, one will lose nothing by believing in him, while if he does exist, one will lose everything by not believing." This reasoning ultimately became known as Pascal's wager.

Blaise Pascal (Courtesy of AIP Emilio Segrè Visual Archives, E. Scott Barr Collection)

Ptolemy, Claudius (ca. 85–168) Known for his astronomy, this Greek mathematician and philosopher wrote the famous series of books known as *The Almagest,* defining the then-known universe mathematically in a series of 13 books. With the original Greek title of *He Mathematike Syntaxis,* or *(The Mathematical Compilation),* the book was eventually translated into Arabic and given the title *Al-Majisti.* The Latin translation of the Arabic title became *Almagest,* and the book became the definitive text on astronomy until Nicolas Copernicus's theory emerged some 1,400 years later. Ptolemy's mathematical computations were geometric and trigonometric in nature, and dealt with devising tables of CHORDS, solving PLANE and spherical TRIANGLES, and using measurements of degrees, minutes, and seconds for ANGLES. He is considered the first person to ever calculate sines.

He approximated the VALUE of pi at 3.1416, calculated the length of the year, and devised the Ptolemy Theorem used in trigonometry, in which a QUADRILATERAL is inscribed within a CIRCLE. He also compiled a star catalogue of 1,028 stars, and wrote several other books, including *Handy Tables,* for the calculation of the positions of heavenly bodies; *Analemna,* which dealt with the mathematics for the construction of a sundial; and *Planetary Hypothesis,* which was written for ordinary people to understand the heavens. He also wrote *Optics,* which discussed reflection and refraction, *Tetrabiblos,* a series of four books on astrology, and *Geography,* a series of eight books on cartography, with the COORDINATES of approximately 8,000 different places defined by longitude and latitude. During the mid-1400s, Latin translations of *Geography* became such prized possessions that owning one became a status symbol. Because Ptolemy created his mathematics to fit to the theories of the day, instead of using mathematics to create original theories based on the facts, many scientists spoke out against his work. The most notable of these was SIR ISAAC NEWTON, who called him a fraud, and declared that Ptolemy's work was "a crime against science and scholarship." Next to *The Elements,* written by EUCLID OF ALEXANDRIA, the *Almagest* is one of the longest-lived textbooks in history.

Pythagoras of Samos (ca. 569–475) A Greek philosopher and leader of a secret society dedicated to religion and mathematics, Pythagoras is often known as the "first pure mathematician," and as founder of a group of followers known as the *mathematikoi,* whose core belief was that all reality was mathematical. Basing their work on defining principles, the Pythagoreans' mathematics did not revolve around solving specific problems, but instead focused on concepts and abstract philosophical mathematical ideas. Pythagoras's contributions are in music theory, number properties, the discovery of IRRATIONAL NUMBERS, and his famous THEOREM for the RIGHT TRIANGLE, the PYTHAGOREAN THEOREM.

Rasiowa, Helena (1917–94) Polish mathematician who studied in the underground version of the University of Warsaw after it was closed by the Nazis in 1939. Rasiowa eventually earned her master's degree in 1946 from the university, and devoted her career to algebraic logic. Her dedicated work in this field led to major contributions in theoretical computer science, and ultimately helped set the stage for scientific work in the field of artificial intelligence.

Recorde, Robert (1510–58) English mathematician famed for giving us the symbol = for "equals." Like many scholars of his time, Recorde studied a variety of topics, including theology, medicine, and mathematics. He was the king's physician, controller of the British mint at Bristol, in charge of the king's silver mines in Ireland, and a prolific writer of textbooks for the masses. Recorde was the first to write mathematics textbooks in English, instead of the traditional languages of Greek or Latin, and he deliberately wrote books that the general public could understand. Some of his books include *The Grounde of Artes,* an arithmetic book that introduces the + for "excess" and – for "deficiency," *Pathwaie to Knowledge,* explaining Euclid's THEOREMS, *The Castle of Knowledge,* which discusses astronomy as taught by PTOLEMY and includes Copernicus's theory of the universe, and *The Whetstone of Witte,* in which he covers such topics as QUADRATIC EQUATIONS and theory, as well as introduces the equals symbol, explaining that "noe 2 thynges can be moare equalle" than two parallel lines.

Regiomontanus (Johann Müller, Johann Müller of Königsberg) (1436–76) German mathematician considered one of the most prominent in his era for the development of the first LOGARITHMic tables of sines, which appeared in an analysis of the *Almagest,* published in 1496. His next work was on astrology, and included another table, this time of tangents, published in 1490. He wrote *De Triangulis* in 1464, which is considered to be the earliest example of PLANE and spherical trigonometry, but this set of five volumes was not published until 1533. All of his trigonometry is known for including the use of algebra. As an early Renaissance man, he helped set the tone for a varied field of interests, and went on to create an observatory, set up a printing press, and invent machines of all sorts. His final work was to revise the calendar, but upon arriving in Rome to begin this task, he died under suspicious circumstances, and it is believed that he was actually murdered.

Rheticus, Georg Joachim von Lauchen (Georg Joachim Iserin, Georg Joachim de Porris) (1514–74) German Renaissance man of science whose fields of study are typical of the time, including mathematics, astronomy, philosophy, medicine, alchemy, and theology. Rheticus was educated by his father, a physician, until the age of 14, at which time his father was beheaded for sorcery, probably related to his medical practice. He received his master's degree from the University of Wittenberg at the age of 22, and immediately began teaching arithmetic and geometry. His love for geometry ultimately led to spending time with Copernicus in 1539, and Rheticus is known as the catalyst for getting Copernicus's work into the world. He obtained funding for publishing *Narratio Prima,* the full title of which was the *First report to Johann Schoner on the Books of the Revolutions of the Learned Gentleman and Distinguished Mathematician, the Reverend Doctor Nicolaus Copernicus of Torun, Canon of Warmia,* and also was responsible for funding the printing of the famous *De Revolutionibus,* to which Rheticus added some trigonometry tables. It was the computation and development of these tables of sines and cosines, using RATIOS in RIGHT TRIANGLEs rather than dealing with lengths of CHORDs in CIRCLEs, as was previously done by CLAUDIUS PTOLEMY and REGIOMONTANUS, which earned Rheticus his place in mathematical history. Six years later, Rheticus took up theology, then returned to medicine and spent the last 20 years of his life as a physician.

Riemann, Bernhard (Georg Friedrich Bernhard Riemann) (1826–66) German mathematician considered one of the most brilliant mathematicians of his time, Riemann was a sickly child educated at home by his father until the age of 10, and remained sick most of his life. His work, mostly in the field of geometry, is credited as leading others to making great discoveries in algebra while they worked on trying to prove his many theories. He is acknowledged most for his original ideas. Among the topics of his work are the general theory of functions of complex VARIABLEs, a famous dissertation on the hypotheses that lie as the foundation of geometry, elliptic function, primes, Abelian functions, and linear differential EQUATIONs. He is considered to have been ahead of his time in

Georg Friedrich Bernhard Riemann (Courtesy of AIP Emilio Segrè Visual Archives, T.J.J. See Collection)

much of his thinking, and is credited for laying the foundation for much of Einstein's work in physics.

Ries, Adam (Adam Riese) (1492–1559) German mathematician credited with writing the first arithmetic book meant to teach the masses rather than exclusively scientists and scholars, as had previously been the practice. His book, *Rechenung nach der lenge, auff den Linihen und Feder,* included the fundamentals of addition, subtraction, MULTIPLICATION, and division, and it incorporated the + and – signs.

Rolle, Michel (1652–1719) French mathematician who believed that differential calculus was "a collection of ingenious fallacies." In 1689 Rolle wrote a book on algebra, *Traité d'algèbre,* which contains a THEOREM that deals with roots of an EQUATION, known as Rolle's theorem. He was a self-taught mathematician who also worked on Euclidean geometry.

Rudolff, Christoff (1499–1545) Polish mathematician whose algebra book, *Die Coss,* is credited as the first German textbook to cover algebra. His most renowned contribution is the invention of the symbol $\sqrt{}$ for SQUARE ROOT.

Sacrobosco *See* JOHN DE HOLYWOOD.

Simpson, Thomas (1710–61) Self-educated British weaver turned mathematician, Simpson became a professor of mathematics and published many works on the sciences, including *Fluxions, Laws of Chance, Annuities and Reversions, Algebra, Geometry, Trigonometry,* several papers dealing with astronomy and physics, and *Miscellaneous Tracts.* He had a reputation for drinking cheap liquor and keeping "low company."

Stevinus, Simon (Simon Stevin) (1548–1620) Dutch military engineer who used symbols to denote EXPONENTs, introduced the use of DECIMALs in mathematics, and suggested using the decimal system for measurements. He published 11 books, including works on geometry, trigonometry, and hydrostatics, the last of which resulted in his fame as the inventor of the field of hydrostatics. Other important work includes his ideas surrounding inclined PLANEs.

Tartaglia, Nicholas (Niccolo Fontana, Niccolo Fontana of Brescia)
(1499–1557) Italian mathematician most noted for his infamous battle with GIROLAMO CARDAN and Cardan's student LODOVICO FERRARI over the origin of the SOLUTION for the CUBIC EQUATION. As a 13-year-old, Tartaglia was a victim of the French army's siege on Brescia, in which his father was murdered in a church, and Tartaglia's skull was split open in three places, his jaw and palate sliced open with a sword, and he was left for dead. Nursed back to life by his mother, he was left with a permanent speech impediment, and became called Tartaglia, which means "the stammerer." Tartaglia taught himself to read and write, and lived in such poverty that in the absence of paper he used tombstones as chalkboards to do his lessons. He eventually became a mathematics teacher in Venice, and wrote several important works, including *Nova Scienze,* dealing with gravity and falling bodies and the efficiency of projectiles, *Inventioni,* on cubic equations, *Trattato de numeri e misuri,* on arithmetic, and a treatise on COEFFICIENTS, numbers, and mercantile arithmetic with algebraic formulas. Tartaglia's scandal occurred when he gave his formula for solving QUARTIC equations to Cardan under the promise that they would never be revealed. Cardan broke the promise, and a public battle ensued between Tartaglia and Ferrari when Tartaglia wrote a book, *New Problems and Inventions,* in which he attacked Cardan. Unfortunately for him, a public debate in 1548 between the two resulted in Tartaglia's humiliation and quick departure, which proved Ferrari as the victor and left Tartaglia disgraced, unable to keep his job, and seriously in debt. Tartaglia was the first to write an Italian translation of Euclid's *Elements,* and is now acknowledged as the cocreator with Cardan of the solution of the cubic equation.

Taussky-Todd, Olga (1906–95) This Austrian mathematician dedicated her career to algebraic number theory and matrix theory. She left Europe in 1938 in an escape from the Nazi regime. Inspired by AMALIE NOETHER to pursue the field of algebraic systems, she wrote nearly 300 papers during her career. Taussky-Todd's work is credited as having inspired research by hundreds of people in matrix theory and computer science. She was a recipient of Austria's highest award, the Cross of Honor, and is considered a pioneer in the field of computer applications.

Theon of Smyrna (The Old Theon, Theon the Mathematician)
(ca. 70–135) Greek mathematician who wrote a textbook on arithmetic that he claimed was an introduction to Plato's mathematics. Instead, the book was more about the interrelation of arithmetic and PRIME NUMBERs, geometry and SQUARES, music, and astronomy. While he includes THEOREMS in this book to help explain these subjects, most agree that he was not terribly familiar with Plato's geometry.

al-Uqlidisi, Abu'l Hasan Ahmad ibn Ibrahim (ca. 920–80) He is believed to have been from Damascus, and his work, translated around 1960, is the earliest example of calculating DECIMAL FRACTIONs, and the earliest use of the equivalent of a decimal point, which was denoted at that time by (′) rather than today's symbol (.).

Viète, François (Franciscus Vieta) (1540–1603) A French lawyer who enjoyed mathematics as a hobby, Viète is known most for his book *In Artem Analyticam Isagoge,* or *Introduction to the Analytical Art,* published in 1591. This book explains how algebra can be used to solve problems in geometry, and gives the basic principles of algebra that are in use today. Viète was the first person to give us a systematic approach to using algebraic notation, specifically using letters to represent known and unknown quantities, and he introduced the term COEFFICIENT. He was the first to suggest using just one notation to represent a quantity of an EXPONENT, instead of using a different letter every time a quantity's exponent changed, and was the first to suggest that PI was infinite. He also wrote on geometry, trigonometry, geography, cosmology, astronomy, and calendar reform.

Von Neumann, John (János van Neumann) (1903–57) Hungarian mathematician who dazzled friends as a child by memorizing the phone book. Von Neumann's brilliance in mathematics was recognized at an early age, but the practicality of earning a living was a concern for his family, and his father did not want him to earn a degree in a subject like mathematics because there was little hope of it paying very well. To abide by his father's wishes, and to follow a path that held his passion, von Neumann enrolled at two universities simultaneously in 1921, the University of Budapest for mathematics and the University

John Von Neumann
(Courtesy of AIP Emilio Segrè Visual Archives)

of Berlin for chemistry. He never attended classes at Budapest, but did the work in his spare time while pursuing his chemistry studies. A move to Zürich caused him to change schools, and at the age of 23 he received his degree in chemical engineering from Zürich's Technische Hochschule. Also in 1926, he received his mathematics doctorate from the University of Budapest. Not surprisingly, he was called a young genius by the mathematicians of this time. In 1933 von Neumann went to Princeton, New Jersey, joining Einstein as one of the original members of the Institute for Advanced Study. Von Neumann's contributions to mathematics include SET theory, measure theory, theory of real VARIABLES, quantum theory, statistical mechanics, quantum mechanics, operator algebras, game theory, and computer science.

Wallis, John (1616–1703) British mathematician considered second in importance in the 17th century only behind Newton. Wallis did not study mathematics in school because it was not considered an important academic topic. When Wallis was 15, his brother introduced him to the subject, and he began to read all of the mathematics books he could find. At the age of 21, in 1638, he graduated with a bachelor of arts degree. He received his master's degree two years later and was ordained, becoming a chaplain. His switch to mathematics came in a strange event, when he was in the right place at the right time and encountered a letter written in code, which he was able to decipher in a couple of hours. He put his newfound talents to work for his country, and quickly became an expert in cryptography. In 1647 he read WILLIAM OUGHTRED's book *Clavis Mathematicae,* and began his passion for mathematics. Within two years, his genius was recognized and he was teaching geometry at Oxford and making major contributions to calculus. In 1655 he published his work on conic sections, and in 1656 he published *Arithmetica Infinitorum (Treatise on Algebra),* which became an instant success. In both works, he uses his new symbol ∞ to represent infinity.

Weierstrass, Karl Theodor Wilhelm (1815–97) A German mathematician known as the father of modern analysis, Weierstrass had an unhappy education filled with conflict over what his father wanted him to study, namely business, and what he wanted to study for himself, mathematics. He barely squeaked through his university studies with any education at all, and with the help of others he began working as a schoolteacher. Weierstrass found this job extremely boring, and spent all of his spare time as a frustrated mathematician, doing calculations in isolation. In 1854 he published a paper on abelian functions (a class of functions named after its discoverer, Niels Abel) that was so brilliant it almost instantly earned him an honorary doctorate degree and many job offers. He landed at the University of Berlin, his job of choice, where he lectured in physics, geometry, mechanics, and calculus. In 1863 he created his theory on REAL NUMBERS, and became famous for his work in analysis.

Widman, Johannes (1462–98) German mathematician who wrote *Mercantile Arithmetic* in 1498, which was the first published introduction of the + and – symbols. Although they meant something slightly different than today's definitions, these symbols were later successfully incorporated into mathematics by ADAM RIES.

Wiles, Andrew John (1953–) British mathematician famous for proving FERMAT'S LAST THEOREM. Wiles read about the THEOREM in a library at the age of 10, and since he was able to understand what it was about, he decided he would solve it himself. Eleven years later, he received his bachelor's degree, then went on to earn his Ph.D., but he did not work on Fermat for his doctorate because, as he recalled, "you could spend years getting nothing." In 1986 he decided to dedicate his life to solving the problem. In 1993 he announced that he had discovered the PROOF, only to find out shortly afterwards that it did not work. After working for 14 more months, with no progress, he was ready to give up on his life's work. But on September 19, 1994, he found it. "Suddenly, totally unexpectedly, I had this incredible revelation. . . . I just stared in disbelief for twenty minutes." He had discovered the SOLUTION to Fermat's Last Theorem, and had solved in his "adult life what had been my childhood dream." For his work, Wiles has received the Royal Medal from the Royal Society of London in 1996, and the American Mathematical Society's Cole Prize in 1997.

Yang Hui (ca. 1238–1298) Chinese government official and mathematician whose books on mathematics are recognized

Karl Theodor Wilhelm Weierstrass (Courtesy of AIP Emilio Segrè Visual Archives)

for their importance in many areas, including a method of studying mathematics that, in the words of a reviewer, "is based on real understanding rather than on rote learning," being the first to show QUADRATIC EQUATIONS with negative COEFFICIENTS, and his work with magic squares in geometry.

Zeno of Elea (ca. 490–25 B.C.E.) Greek philosopher whose four paradoxes on motion, as described by Aristotle, relate to the field of calculus and infinitesimals. The ideas state that if something has a magnitude and it can be divided, then it can be divided infinitely; and if it has no magnitude, then it does not exist. Zeno's Dichotomy states "There is no motion, because that which is moved must arrive at the middle before it arrives at the end, and so on ad infinitum." The Achilles is, "The slower will never be overtaken by the quicker, for that which is pursuing must first reach the point from which that which is fleeing started, so that the slower must always be some DISTANCE ahead." The fable of the tortoise and the hare embodies this idea. His paradox called the Arrow is, "If everything is either at rest or moving when it occupies a space equal to itself, while the object moved is always in the instant, a moving arrow is unmoved." And the Stadium states, "Consider two rows of bodies, each composed of an equal number of bodies of equal size. They pass each other as they travel with equal VELOCITY in opposite directions. Thus, half a time is equal to the whole time."

Zhang Heng (78–139) Chinese astrologer, astronomer, geographer, and mathematician who invented the first seismograph, which he called the "instrument for inquiring into the wind and the shaking of the earth." He also constructed the first rotating globe of the Earth in China, which demonstrated his belief that the world was round.

SECTION THREE
CHRONOLOGY

2000 B.C.E. ● The Babylonians are the first to solve QUADRATIC EQUATIONS in radicals.

ca. 1650 B.C.E. ● The Ahmes papyrus is written, and contains exercises in arithmetic and geometry, including FRACTIONS, notation, linear algebra, ISOSCELES TRIANGLES and TRAPEZOIDS, QUADRILATERALS, and AREAS. The ancient Egyptians determine that the value of pi is 3.1605.

ca. 700 B.C.E. ● The Babylonians identify the 12 signs of the Zodiac.

ca. 670 B.C.E. ● Through their mathematical computations, the Babylonians are successful in predicting solar eclipses.

ca. 600 B.C.E. ● One of the seven sages of Greece, Thales, is founder of the earliest known school of philosophy and mathematics.

ca. 569 B.C.E. ● PYTHAGORAS OF SAMOS is born.

ca. 549 B.C.E. ● Pythagoras starts a secret philosophical and mathematical society that includes both men and women as members who are sworn to secrecy. They are known as Pythagoreans.

ca. 480 B.C.E. ● The ancient Greek philosopher Oenopides calculates that the Earth is a sphere that is tipped 24 degrees from its plane of orbit.

ca. 475 B.C.E. ● A mob attacks a group of members of the secret Pythagorean society, and murders Pythagoras.

ca. 470 B.C.E. ● Hippasus, a Pythagorean, violates his oath of secrecy and is drowned.

ca. 450 B.C.E. ● Philolaus, a Greek philosopher in the Pythagorean society, believes that the Earth rotates.

ca. 370 B.C.E. ● Philolaus writes the first book about the Pythagoreans. By this time, there is no longer a threat of death for discussing the society's beliefs.

ca. 350 B.C.E. ● Aristotle theorizes that the Earth is the center of the universe, around which everything else revolves.

ca. 300 B.C.E. ● EUCLID writes *The Elements.*

ca. 250 B.C.E. ● ARISTARCHUS OF SAMOS uses mathematical computations to determine that the Sun, rather than the Earth, is the center of the universe. For this he is often called the Ancient Copernicus.

● The Greek philosopher Eratosthenes calculates the Earth's circumference to be 28,500 miles.

ca. 212 B.C.E. ● After a three-year siege on Syracuse, and despite orders to the contrary, a Roman soldier kills ARCHIMEDES OF SYRACUSE, one of the greatest mathematicians of all times. Archimedes' tomb is inscribed with a SPHERE and a cylinder, per his wishes, to signify his favorite work showing that the VOLUME of a sphere and the surface of a sphere, are two-thirds the volume and surface of a circumscribed cylinder.

ca. 200 B.C.E. ● APOLLONIUS OF PERGA gives the approximation of pi as 3.1416, although he gives no explanation of how he arrives at this VALUE.

ca. 150 B.C.E. ● Greek astronomer Hipparchus uses longitude and latitude measurements to determine specific locations of geographical points.

ca. 130 B.C.E. ● Greek astronomer Hipparchus calculates the size of the Moon and its distance from Earth.

ca. 100 C.E. ● NICOMACHUS OF GERASA writes a book on arithmetic explaining properties of numbers, FRACTIONS, and RATIOS.

ca. 140 ● CLAUDIUS PTOLEMY becomes the first person to devise calculations that are comparative to sines. He writes *The Almagest,* which becomes the basis of astronomy teaching for the next 1,400 years, ranking the book as the second-longest-running textbook in history.

ca. 175 ● Gerasa's *Introduction to Arithmetic* is translated into Latin, and remains the primary arithmetic textbook for the next 1,000 years.

ca. 250 ● DIOPHANTUS OF ALEXANDRIA is the first to use symbols to represent unknown quantities.

400 ● HYPATIA becomes head of the Platonist school in Alexandria, where her lectures include philosophy and mathematics.

415 ● A mob of Christians attack and murder Hypatia for her scientific and philosophical ideas because they are thought to be pagan.

ca. 500 ● A Greek anthology is compiled that contains the now-famous puzzle: Diophantus spent $\frac{1}{6}$ of his life in childhood, $\frac{1}{12}$ in youth, $\frac{1}{7}$ more as a bachelor; five years after his marriage, his son was born, who died four years before Diophantus at half his age, so how long did Diophantus live?

665 ● BRAHMAGUPTA invents the concept of ZERO.

ca. 800 ● Arabian ABU JA'FAR MUHAMMAD IBN MUSA AL-KHWĀRIZMĪ writes one of the most important mathematical texts of all time, using al-gebr to work through real-life problems with calculations of NATURAL NUMBERS, EQUATIONS, and FRACTIONS. This is the first recorded work to introduce the system of base 10 numbers.

850 ● The *Ganita Sera Samgraha* is written by MAHAVIRA as an expansion of Brahmagupta's work, and becomes the first Indian book devoted solely to mathematics.

1027 ● ABU ALI AL-HASAN IBN AL-HAYTHAM writes his autobiography.

1079 ● OMAR KHAYYAM works on reforming the calendar, and gives the length of the year as 365.24219858156 days, remarkably close to the length of 365.242190 known today.

1100 ● Al-Khwārizmī's work, *Al-gebr*, is translated into Latin, giving us the term *algebra*.

1202 ● LEONARDO PISANO FIBONACCI writes *Liber abbaci,* his first book on mathematics, which is responsible for changing the calculating methods of Europe from the ROMAN NUMERAL system to Arabic numbers. This book also is the first introduction of the FIBONACCI SERIES, which is explained in the book as "A certain man put a pair of rabbits in a place surrounded on all sides by a wall. How many pairs of rabbits can be produced from that pair in a year if it is supposed that every month each pair begets a new pair which from the second month on becomes productive?"

1220 ● Mathematician JOHN DE HOLYWOOD writes *Tractatus de Sphera* as an astronomy text, and it is used as the basic text on the subject for the next four centuries.

1232 ● John de Holywood writes the book *De Anni Ratione* on the study of time and the calendar, and states that the way to correct the Julian calendar is to remove one day from the calendar every 288 years.

1406 ● Jacopo Angeli da Scarperia translates Ptolemy's *Geography* into Latin. Within the next 40 years, owning a copy of this book, with its beautiful illustrations, becomes a status symbol among the wealthy.

1450 ● Johannes Gutenberg develops the first commercial printing press, marking the beginning of the end of handwritten books, and making it possible for the masses to acquire knowledge at an affordable price.

1463 ● REGIOMONTANUS cowrites a book with Georg Peurbach entitled the *Epitome of the Almagest,* and it becomes a leading text for scholars of astronomy for the next 150 years.

1482 ● Euclid's book, *The Elements,* is published for the first time. It will become the most translated and published textbook of all time, going through more than 1,000 editions.

1494 ● LUCAS PACIOLI publishes a 600-page summary of mathematics, *Summa de Arithmetica, Geometria, Proportioni et Proportionalita,* which covers everything known about the topic at the time, and is considered the first printed book on all of these subjects.

1497 ● Pacioli starts work on his second book, *Divina Proportione,* with the help of his friend Leonardo da Vinci as illustrator, and the work turns into a three-volume compilation that is finally published in 1509.

1498 ● JOHANNES WIDMAN writes *Mercantile Arithmetic,* in which the symbols of + and – are introduced in a book for the first time.

1499 ● Pacioli and da Vinci flee the politically volatile city of Milan, ultimately ending up in Florence, where they become housemates for the next seven years.

1525 ● CHRISTOFF RUDOLFF invents the symbol for SQUARE ROOT, which is published in his algebra book, *Die Coss.*

1533 ● Dutch geographer Reiner Dokkum suggests that longitude can be determined by using the position of the Sun and a clock.

1539 ● Mathematician GEORG JOACHIM VON LAUCHEN RHETICUS visits Nicolaus Copernicus, and raises funds to publish his *Narratio Prima,* which ultimately gains a reputation as the best introduction to Copernicus's work.

1540 ● ROBERT RECORDE publishes an arithmetic book, *Grounde of Artes,* which is the first to use the plus sign to signify an excess and the minus sign to signify a deficiency.

1541 ● Rheticus convinces Duke Albert of Prussia to fund the printing of Copernicus's *De Revolutionibus,* and elaborates the book with his own trigonometry tables. These are the first published tables to include cosines, and Rheticus's contribution to mathematics is considered to be of astronomical value.

1545 ● Ferrari's solution of the QUARTIC EQUATION is published in Girolamo Cardan's book, *The Great Art, or the Rules of Algebra,* but since the world is three-dimensional and Ferrari's work dealt with taking a quantity to the fourth power, his work is thought to be ridiculous.

1557 ● Robert Recorde introduces the EQUALS SIGN, =, in his book *Whetstone of Witte,* based on his belief that nothing is more equal than PARALLEL lines. Nonetheless, mathematicians continue to write the word, instead of using a symbol, until about 1600. The title of Recorde's book is a pun, based on the Latin word *cosa,* used to represent unknown quantities in algebra, *Die Coss,* the name of Christoff Rudolff's algebra book, and *cos,* which is Latin for "whetstone." In this case, Recorde suggests that the mathematics help sharpen the wit.

1570 ● RAFAELLO BOMBELLI translates much of Diophantus's *Arithmetica* from Greek into Latin, but never publishes the work.

1572 ● Rafaello Bombelli is the first to suggest using symbols to denote EXPONENTS.

1583 ● Thomas Finck of Denmark invents the terms SECANT and TANGENT.

1586 ● SIMON STEVINUS states that objects of different weights fall at the same rate, which is stated three years later by GALILEO GALILEI.

1591 ● FRANÇOISE VIÈTE writes *In Artem Analyticam Isagoge,* or *Introduction to the Analytical Art,* establishing the basic principles of algebra and leading to modern algebra notation.

1593 ● Galileo Galilei invents the first thermometer, a water and air thermometer, to measure temperature.

1600 ● Mathematicians begin to phase out the written word *aequalis,* which we now write as EQUALS, replacing it with a symbol. Consistent use of the equals symbol, =, as used today, does not occur until some 80 years later.

1609 ● Galileo is appointed court mathematician and philosopher in Florence by the Grand Duke of Tuscany.

1610 ● THOMAS HARRIOT discovers sunspots and deduces the Sun's rotation based on his observations.

● After nearly a lifetime of work, German mathematician Ludolph Van Ceulen calculates the value of pi to 35 places through the use of POLYGONs with 2^{62} sides. As a result of this phenomenal work, pi is often referred to as the Ludolphine number.

1613 ● PIETRO ANTONIO CATALDI invents CONTINUED FRACTIONS.

1614 ● JOHN NAPIER invents the concept of LOGARITHMS, discussed in his book *Mirifici Logarithmorum Canonis Descriptio,* which is translated from Latin into English two years later. In 1615 he visits HENRY BRIGGS and encourages him to develop base 10 logarithmic tables, using log 1 = 0.

1617 ● Napier publishes his book *Rabdologia,* in which he introduces his mechanical system for multiplying and dividing, known as Napier's bones.

1619 ● RENÉ DESCARTES has a series of dreams about philosophy and analytical geometry on the night of November 10, causing him to change the focus of his life's work.

1621 ● French writer Claude-Gaspard Bachet de Méziriac translates and publishes Diophantus's *Arithmetica* from Greek into Latin. Bachet's becomes the most famous translation of the work.

1624 ● Claude-Gaspard Bachet de Méziriac's second edition of his book *Problèmes plaisants* includes mathematical tricks, and he is the first to suggest CONTINUED FRACTIONS as a solution of indeterminate equations.

1629 ● BONAVENTURA CAVALIERI is the first to state the principle of indivisibles, in which a line, a surface, and a VOLUME are each made up of an infinite number of points, lines, and surfaces, respectively. This was first printed in 1635.

● Albert Girard is the first to use brackets in mathematics.

1631 ● Harriot's book *Aequationes Algebraicas Resolvendas* is published 10 years after his death, and is the first to introduce the symbols of GREATER THAN and LESS THAN.

● Harriot uses a dot as the symbol for MULTIPLICATION.

● WILLIAM OUGHTRED publishes *Clavis Mathematicae,* introducing the symbol x for multiplication.

1632 ● Galileo writes *Dialog Concerning the Two Chief World Systems,* which is not an explicit statement that he does not believe in Aristotle's and Ptolemy's teachings, but is an obvious attempt at putting forth the idea that they are wrong. This work lands him in trouble with the Roman Catholic Church for the last time.

1633 ● Galileo is hauled before the Inquisition to recant his beliefs and save himself from death for heresy. He is placed under house arrest for life.

● Descartes finishes four years of work on his book *Le Monde,* which explains his theory of the universe. After hearing about Galileo's house arrest, he decides not to publish the work. It is eventually published in 1664, 14 years after his death.

1635 ● At the age of 12, BLAISE PASCAL receives a copy of his first mathematics book, Euclid's *Elements.*

1637 ● Descartes publishes his most important contribution to mathematics, *La Géométrie,* crediting him as the inventor of analytical geometry.

● Descartes uses juxtaposition to show multiplication.

1639 ● Pascal writes his first geometry essay at the age of 16, containing his THEORY on conics that gives us Pascal's Theorem. The essay remains unprinted for more than a century, and is finally published in 1779.

1641 ● Publication of *Meditationes* by Descartes makes him the founder of modern philosophy.

● The 18-year-old Pascal invents the world's first digital calculator, which he devised to help his father with tax collections. Only 50 of the machines will be built by 1652, and no more will be produced. A similar design will not appear until 1940.

1642 ● Galileo dies on Christmas Day. This same day, SIR ISAAC NEWTON is born.

1647 ● Descartes's theory of vortices, found in his book *Principia Philosophiae,* becomes a major contribution to the field of mathematics.

1652 ● Newton starts elementary school at the age of 10. He is soon pulled out of school because he doesn't seem to have an affinity for learning, and is not allowed to return until he is 13 years old.

1653 ● PIERRE DE FERMAT is mistakenly reported as having died from the plague.

● Pascal writes *Treatise on the Equilibrium of Liquids,* giving us Pascal's law of pressure, considered a major contribution to physics.

● Descartes's paper on music theory, *Renati Descartes Musicae Compendium,* published in 1650 by a Dutch publisher, is translated by mathematician VISCOUNT WILLIAM BROUNCKER and expanded to twice the size of the original work. This becomes Brouncker's only published book.

1654 ● Mathematician Blaise Pascal begins a correspondence with Fermat about the PROBABILITY of rolling a double six on dice. Over the course of the summer, exchanging five letters between them, they create the theory of probability.

1655 ● JOHN WALLIS introduces the symbol ∞ for infinity, which had previously been used by the Romans as the number 1,000. The symbol does not appear to be used again for some 60 years, until JAKOB BERNOULLI uses it in his book *Ars Conjectandi.*

1656 ● Wallis writes *Arithmetica Infinitorum,* which instantly becomes the definitive book on arithmetic.

1657 ● Frans van Schooten recommends using COORDINATES in three-dimensional space.

1658 ● Pascal, having turned to religion and discarding his study of mathematics, writes his famous work in philosophy on life and death, *Pensées,* and creates Pascal's wager.

1659 ● Vincenzo Viviani, a student of Galileo's, is responsible for the discovery and restoration of a lost book on conic sections written by Apollonius.

1661 ● Newton enters Trinity College, Cambridge, at the age of 18. He receives financial aid from the school, and in exchange is expected to work as a servant for the wealthier students.

1662 ● With the help of his connection to King Charles II, VISCOUNT WILLIAM BROUNCKER helps cofound the Royal Society of London, and is elected as its first president.

1663 ● JAMES GREGORY publishes his book, *Optica Promota,* in which he gives the mathematical computations for his new invention, the first reflecting telescope.

● Newton borrows a copy of Euclid's *Elements* to try to grasp an understanding of the computations in an astronomy book. This is his first real look at mathematics. He is 20 years old.

1665 ● The plague forces Cambridge University to close, and Newton returns to his family's home. In the following year and a half, he will make discoveries that change the nature of physics, optics, astronomy, and mathematics.

1671 ● Newton writes *De Methodis Serierum et Floxionum,* his work on calculus. He does nothing to get it published, and it will remain virtually unknown until it is translated into English and printed in 1736. Newton's slowness to publish causes many of the problems that later plague him and GOTTFRIED WILHELM LEIBNIZ in their dispute over who invented calculus.

1672 ● John Pell publishes the first table of SQUARE numbers.

● Newton donates a reflecting telescope to the Royal Society of London. They elect him as a fellow into the society, and publish his article on optics.

1675 ● Mathematician and inventor Robert Hooke has a balance spring watch made under his direction, based on his mathematical discoveries of oscillations of a coiled spring.

Rival mathematician Huygens had done the same three months earlier, unbeknownst to Hooke.

- The speed of light is first measured by Ole Roemer. He also uses his mathematical mind to figure out the best shape for the teeth of a geared wheel to provide the smoothest motion.

- Hooke accuses Newton of stealing his optics research, and the Jesuits attack Newton's optics theories on light and color.

1678 ● Newton suffers his first nervous breakdown.

ca. 1680 ● The symbol commonly used for equals between the years 1600 and 1680, which resembles today's infinity sign, is replaced with the symbol =, introduced by Recorde more than a century earlier.

1687 ● With urging from Edmond Halley, Newton publishes *Principia,* considered one of the greatest scientific books of all time. In it, Newton explains his Universal Law of Gravitation and his Laws of Motion.

1693 ● Newton suffers his second nervous breakdown, and leaves the fields of mathematics, physics, and astronomy research forever. Three years later, he will take a government job running the Royal Mint, and become a very wealthy man.

1696 ● JOHANN BERNOULLI's student, Guillaume de l'Hopital, uses his teacher's lessons without permission as the basis for a book, and publishes *Analyse des infiniment petits pour l'intelligence des lignes courbes,* the first textbook ever written on calculus.

1703 ● The Royal Society of London elects Newton as its president.

1704 ● Newton finally publishes his entire optical research, entitled *Optiks,* the year after Hooke's death.

1705 ● Newton is knighted by Queen Anne for his contributions to science.

1711 ● The Royal Society publishes an article claiming Leibniz, who is considered the inventor of calculus, plagiarized

Newton's work in the field. A bitter dispute follows. As president of the organization, Newton requests an investigation, then anonymously writes the society's official stance stating that Newton is the true inventor.

1713 ● JAKOB BERNOULLI's book *Ars conjectandi* is published, eight years after his death, stating the properties of the Bernoulli numbers.

● ROGER COTES completes his work editing Newton's second edition of *Principia*.

1714 ● GABRIEL DANIEL FAHRENHEIT invents the first mercury thermometer.

1724 ● Fahrenheit introduces a new temperature scale to measure the freezing and boiling points of liquids in his mercury thermometer, called the Fahrenheit scale.

1727 ● Euler introduces the symbol *e* to represent the BASE of natural logs.

1734 ● French scientist Pierre Bouguer introduces the symbols for "greater than or equal to" and "less than or equal to."

● Euler invents *f(x)* as the notation of a function.

1742 ● GABRIEL CRAMER publishes Johann Bernoulli's *Complete Works* in a four-volume set. Bernoulli insisted that the work could only be published by Cramer.

● ANDERS CELSIUS develops the centigrade scale, also called the Celsius scale, to measure temperature.

● CHRISTIAN GOLDBACH writes a letter to Euler in which he states his CONJECTURE that every even INTEGER greater than two is the SUM of two primes. This becomes known as Goldbach's Conjecture, and will remain one of the greatest unproved mathematical problems of all time.

1744 ● Jakob Bernoulli's *Works* is published by Gabriel Cramer. This two-volume set includes material that was previously unknown, but does not include the previously published *Ars conjectandi.*

1755 ● Euler introduces the Greek letter sigma, notated as Σ and called the SIGMA NOTATION, as the symbol for summation.

● Gabriel Cramer publishes his famous book of algebraic curves modeled from Newton's memoirs, entitled *Introduction à l'analyse des lignes courbes algébraique.*

1766 ● Euler coins the term "calculus of variations."

1771 ● Euler's house burns down, and he loses much of his work, which he later rewrites with improvements. Shortly after the fire, Euler undergoes a cataract operation, and within a few days goes completely blind. Half of his published work is written over the next 12 years.

1777 ● Euler designates i as the $\sqrt{-1}$.

1785 ● PIERRE-SIMON LAPLACE gives his 16-year-old mathematics student, Napoleon Bonaparte, a passing grade.

1794 ● MARIE-SOPHIE GERMAIN assumes the identity of Monsieur Le Blanc, a former student at the college she was unable to attend as a woman, and begins her incognito studies of mathematics with Lagrange. He eventually will discover her true identity and inspire her to begin work on FERMAT'S LAST THEOREM, which becomes her biggest contribution to mathematics.

1797 ● CARL FRIEDRICH GAUSS devises the PROOF for the FUNDAMENTAL THEOREM OF ALGEBRA, putting an end to the 250-year-long search for a formula to solve the QUINTIC, or fifth degree, equation. He is 20 years old.

1798 ● Gauss constructs a 17-sided POLYGON, the HEPTADECAGON, using just a straightedge and a compass.

1799 ● A slab of basalt rock is found, with writings carved into it in three languages: Egyptian hieroglyphic, Egyptian demotic, and Greek. Known as the Rosetta Stone, this rock will make it possible to decipher the Ahmes papyrus, discovered some 100 years later, and written nearly 4,000 years earlier.

1799 ● The Royal Danish Academy publishes a paper by Caspar Wessel that shows the geometric interpretation of complex numbers, but the work will go unnoticed until 1895. Wessel will ultimately share credit with JEAN ROBERT ARGAND for this work.

1801 ● Gauss publishes his first book, *Disquisitiones Arithmeticae,* devoted primarily to number theory.

1806 ● Amateur mathematician Jean Robert Argand is the first to give a PROOF on the FUNDAMENTAL THEOREM OF ALGEBRA in which the COEFFICIENTS are COMPLEX NUMBERS. Argand also self-publishes a text on the geometric interpretation of complex numbers, but fails to put his name on the book. The work will remain anonymous until 1813, and eventually becomes known as Argand's diagram.

1812 ● CHARLES BABBAGE, the "father of computing," makes the first mechanical calculator.

1823 ● The British government begins funding the first digital computer, known as the Difference Engine, designed by Babbage. Funds run out by 1827, and friends of Babbage manage to get more funding from the Duke of Wellington in 1829. In 1834, all of the money is gone again, and Babbage asks the government for more. For the next eight years, the British government refuses to give Babbage an answer as to whether they will give him more money, and in 1842 they finally tell him they are officially not going to fund the project. The machine remains unfinished during Babbage's lifetime, and Babbage is ridiculed for wasting so much money.

1825 ● Dirichlet and Legendre prove Fermat's Last Theorem for $n = 5$.

1832 ● EVARISTE GALOIS writes his idea on group theory, and is shot the next day in a duel that ends his life at the age of 20.

1837 ● Dirichlet defines a function: "If a variable y is so related to a variable x that whenever a numerical value is assigned to x, there is a rule according to which a unique value of y is determined, then y is said to be a function of the independent variable x."

1844 ● A 20-year-old German named Johann Dase correctly computes the value of pi in his head to 205 decimal places, without writing any of it down.

1848 ● Sir William Thomson Kelvin creates the Kelvin scale to measure absolute temperature.

1850 ● Amadee Mannheim, an officer of the French army, invents a new design for the slide rule, which will remain the standard design.

1854 ● GEORGE BOOLE publishes *An Investigation into the Laws of Thought, on Which Are Founded the Mathematical Theories of Logic and Probabilities,* introducing the world to the algebra of logic. Called BOOLEAN ALGEBRA, it will make the invention of telephone switches and computers possible.

1855 ● FLORENCE NIGHTINGALE develops the Polar Area Diagram, a GRAPH she calls coxcombs, to show the mortality rates of British soldiers based on statistics she gathers, and she computes that the entire British army will be killed by disease if hospital sanitation does not improve.

1857 ● England establishes the Royal Commission on Health for the Army based on Nightingale's statistics for sanitary hospital reform. She will go on to advise Canada and the United States on military medical care and ultimately reshape the nature of the nursing profession.

1858 ● A papyrus is found in a ruin in Thebes, and is purchased by Henry Rhind. It is the Ahmes papyrus (also called the Rhind Papyrus) written nearly 4,000 years earlier by the scribe Ahmes, and contains more than 80 mathematical exercises.

    ● On November 24, RICHARD DEDEKIND comes up with the concept of the Dedekind cut.

    ● Florence Nightingale becomes the first woman Fellow of the Royal Statistical Society.

1859 ● Edward Fitzgerald translates nearly 600 poems. Known as the *Rubaiyat,* these poems are attributed to mathematician OMAR KHAYYAM.

1864 ● SOFIA VASILYEVNA KOVALEVSKAYA teaches herself trigonometry at the age of 14 in order to understand a physics book written by her neighbor and given to her family as a gift.

1866 ● Sir Thomas Clifford Allbutt invents the first "short" medical thermometer, taking the place of the device previously used, which was a foot long and took 20 minutes to measure a temperature.

1873 ● William Shanks, owner of a British boarding school, calculates the value of pi to 707 decimal places by hand. In 1944, with the aid of a calculator, it will be discovered that Shanks made a mistake at the 526th entry.

1876 ● FRANÇOIS-EDOUARD-ANATOLE LUCAS discovers that the Mersenne number $2^{172} - 1$ is a PRIME NUMBER, the largest prime number ever calculated without the use of a computer.

1880 ● M. Pépin solves the system of equations in integers, $x^2 + y^2 = z^2$, $x^2 = u^2 + v^2$, $x - y = u - v$, given to Huygens as a challenge from French mathematician Bernard Frénicle de Bessy some 200 years earlier.

1883 ● Lucas invents the mathematical puzzle called the Tower of Hanoi, and publishes it under the name of Claus, which is an anagram for Lucas.

1886 ● Kovalevskaya wins the Prix Bordin from the French Academy of Sciences for her paper on the rotation of a solid body about a fixed point, which is so impressive that the prize money is increased from 3,000 to 5,000 francs.

1921 ● AMELIE NOETHER publishes her book, *Idealtheorie in Ringbereichen,* which is credited as a major development in the advancement of modern algebra.

1935 ● Albert Einstein writes Noether's obituary, and lauds her as a mathematical genius.

1938 ● Dame MARY CARTWRIGHT and John Littlewood begin work for the Radio Research Board of the Department of Scientific and Industrial Research. Through their efforts the chaos theory is born.

1960 ● An obscure Arabic text by A1 Uqlidisi is rediscovered and translated, and is found to include the earliest examples of a DECIMAL place, dating from ca. 950.

1963 ● ANDREW JOHN WILES finds a book on Fermat's Last Theorem in his local library, and he decides that he is going to solve the theorem. He is 10 years old.

1968 ● An Arabic manuscript is found in the Astan-I Quds library in Iran by F. Sezgin, which proves to be a translation of the lost Books IV to VII of Diophantus's *Arithmetic.*

1975 ● TJALLING CHARLES KOOPMANS and Leonid Kantrovich share the Nobel Prize in economics for their real-world application of mathematics in scheduling and allocating resources, which is known as linear programming.

1979 ● The Roman Catholic Church, under the orders of Pope John Paul II, opens the file on Galileo to consider reversal of his condemnation. The church will officially acknowledge its mistake in 1992.

1990 ● Japanese mathematician SHIGEFUMI MORI wins the prestigious Fields Medal and the Cole Prize in Algebra for his work on algebraic geometry.

1993 ● Wiles claims to have proved Fermat's Last Theorem, and later in the year recants his claim due to problems with the proof.

1994 ● Wiles once again claims that he has proved Fermat's Last Theorem. This time, he is right, solving a problem that has plagued mathematicians for 350 years.

1998 ● With the use of computers, Goldbach's Conjecture was shown to hold true in calculations up to 400,000,000,000,000.

Nonetheless, a proof for the CONJECTURE has never been discovered.

● A mathematician from the University of Michigan, Thomas Hales, uses computer-generated data to prove THOMAS HARRIOT's conjecture on stacking spheres.

SECTION FOUR
CHARTS & TABLES

Absolute Value

The absolute value of a number is the number itself, without a positive or negative sign in front of it. For example, the absolute value of –3 is 3, and the absolute value of +5 is 5. Instead of writing out the words, "the absolute value of" a quantity, the symbol for absolute value | | is used to express the same idea, like this: |–3|.

To solve for the absolute value of a number, the answer is the number itself. In other words, it is the number without the positive or negative sign. Here are some examples:

$|-3| = 3$
$|+3| = 3$
$|3| = 3$
$|+5| = 5$
$|-72| = 72$
$|-9| = 9$
$|0| = 0$

If a number does not have a sign in front of it, it is considered a positive number, as in the example |3| above.

Zero is neither a negative nor a positive number, but the absolute value of 0 is still 0.

Area

Area is the surface space of a figure.
Here is how to calculate area *(A)* for a variety of figures:

Circle
  Area = $\pi \times$ the square of the radius
  $A = \pi r^2$

Parallelogram
  Area = base × height
  $A = bh$

Rectangle/Square
  Area = length × width
  $A = lw$

(continues)

Area *(continued)*

Rhombus

Area = (length of diagonal 1 × length of diagonal 2) ÷ 2

$A = d_1 d_2 \div 2$

Surface of a sphere

Area = π × square of the diameter

$A = \pi d^2$

Surface of a cylinder

Area = 2 × π × square of the radius + 2 × π × the radius × the height

$A = 2\pi r^2 + 2\pi rh$

Trapezoid

Area = height × length of each parallel side a and b ÷ 2

$A = h(a + b) \div 2$

Triangle

Area = base × height ÷ 2

$A = bh \div 2$

Arithmetic Sequence and Common Difference

An arithmetic sequence is a set of numbers or terms that has a pattern running through it, and the pattern has a common difference. This means that when one number or term in an arithmetic sequence is subtracted from the number or term that immediately follows it, the result is always the same: it is the common difference. Here are some arithmetic sequences, and their common differences:

Arithmetic Sequence	Common Difference
2, 4, 6, 8, 10, _____	2
$4 - 2 = 2$	
$6 - 4 = 2$	
$8 - 6 = 2$	
$10 - 8 = 2$	

(continues)

Arithmetic Sequence and Common Difference *(continued)*

Arithmetic Sequence	Common Difference
1, 3, 5, 7, 9, _____	2

$$3 - 1 = 2$$
$$5 - 3 = 2$$
$$7 - 5 = 2$$
$$9 - 7 = 2$$

3, 6, 9, 12, 15, _____	3

$$6 - 3 = 3$$
$$9 - 6 = 3$$
$$12 - 9 = 3$$
$$15 - 12 = 3$$

1, 4, 7, 10, 13, _____	3

$$4 - 1 = 3$$
$$7 - 4 = 3$$
$$10 - 7 = 3$$
$$13 - 10 = 3$$

1, 5, 9, 13, 17, _____	4

$$5 - 1 = 4$$
$$9 - 5 = 4$$
$$13 - 9 = 4$$
$$17 - 13 = 4$$

5, 10, 15, 20, 25, _____	5

$$10 - 5 = 5$$
$$15 - 10 = 5$$
$$20 - 15 = 5$$
$$25 - 20 = 5$$

2, 8, 14, 20, 26, _____	6

$$8 - 2 = 6$$
$$14 - 8 = 6$$
$$20 - 14 = 6$$
$$26 - 20 = 6$$

Boolean Algebra

This particular type of algebra is based on logic and logical statements, and is used for sets, diagrams, probability, and computer design and applications. Typically, statements are represented by letters such as *p, q, r,* and *s.* For example:

>   *p* is "3 belongs to a set of odd numbers"
>   *q* is "2 belongs to the same set of numbers as 3"
>   *r* is "ice cream is a dairy product"
>   *s* is "chocolate ice cream is the best-selling dairy product"

Statements can be true or false, and they can be combined to form compound statements. Since compound statements in Boolean algebra follow logic, one statement can be true, another can be false, and when combined together the compound statement can be either true or false depending on the conditions placed on the statement. Let's use *p* and *q* for our example compound statements:

To indicate	it is written	and it is called a(n)
p and *q*	$p \wedge q$	conjunction
p or *q*	$p \vee q$	disjunction
if *p*, then *q*	$p \rightarrow q$	conditional
p if and only if *q*	$p \leftrightarrow q$	equivalence
not *p*	p'	negation

The "truths" of the compound statements are determined by the value of each statement on its own, and then the combined value. So, in the example of the conjunction *p* and *q*, written $p \wedge q$, if *p* is true on its own and *q* is true on its own, then the conjunction of *p* and *q* is true. But if either *p* or *q* is false, then the conjunction statement $p \wedge q$ is false.

In a disjunction, however, where the statement is *p* or *q*, written $p \vee q$, if one of the statements is true then the compound statement is true, and it is only false if both statements are false.

Cartesian Coordinate System

The standard graph used to plot points of ordered pairs is the Cartesian coordinate system. It is made up of two axes, and has four quadrants. Here are some of the elements of this system:

> The *x-axis* is the horizontal axis.
> The *y-axis* is the vertical axis.

An *x* point on the graph is called the abscissa.

A *y* point on the graph is called the ordinate.

When an *x* point and a *y* point are indicated together, they become coordinates on the graph. Written together as *(x,y),* they are called an ordered pair.

The *ordered pairs* indicate where the numbers are to be plotted on the graph, and the order in which they are written is significant. The first number corresponds with the point on the *x*-axis, and the second number corresponds with the point on the *y*-axis, *(x,y).*

A *function* is a set of ordered pairs in which all of the *x*-coordinate numbers are different.

> This is not a function, because not all of the *x*-coordinates are different:
> (3, 5), (5, 2) (5, 6), (8, 3)
> This is a function, because all of the *x*-coordinates are different:
> (2, 7), (4, 3), (7, 4), (9, 6).

The *origin* is the point of intersection of the *x*-axis and the *y*-axis on the graph. It is indicated with the letter *O,* and has the ordered pair (0, 0).

A *quadrant* is one of the four sections of the graph used for plotting coordinates. The quadrants are the result of the interception of the *x*-axis and *y*-axis.

Celsius and Fahrenheit Conversion

From the Celsius system to the Fahrenheit system of temperature, or vice versa, the calculation is:

Fahrenheit = $(C \times 1.8) + 32°$

Celsius = $(F - 32°) \div 1.8$

Decimals

The word *decimals* usually refers to *decimal fractions* in the base 10 number system. The dot, or decimal point, is used to show both the integer and the fraction values. The numbers to the left of the dot are the integers, and the numbers to the right of the dot are the fractions. For example, 0.4, 3.6, 1.85, 97.029, and so on.

Each place, or order, to the right of the decimal point has a name to identify the value of the fraction. For example, 0.4 is read as four tenths. 0.62 is read as sixty-two hundredths. 0.759 is read as seven hundred fifty-nine thousandths. 3.6 is read as three and six tenths. 7.759 is read as seven and seven hundred fifty-nine thousandths.

Here are the values for the first six places in decimal fractions:

.1	Tenths	(one decimal place)
.01	Hundredths	(two decimal places)
.001	Thousandths	(three decimal places)
.0001	Ten-thousandths	(four decimal places)
.00001	Hundred-thousandths	(five decimal places)
.000001	Millionths	(six decimal places)

Degrees of a Term

In expressions and equations, numbers and variables have degrees of terms, based on the power to which a number or variable is raised. A number by itself is said to have a zero degree; a variable raised to the second power (squared) is said to be a second-degree term. Here are some degrees of terms:

3	zero degree
x	first degree
x^2	second degree
x^3	third degree
x^4	fourth degree
x^5	fifth degree

Exponents

Fractional Exponents
$$x^{n/b} = (^b\sqrt{x})^n = {^b\sqrt{x^n}}$$

Division
$$b^m/b^n = b^{m-n}$$
or
$$= 1/b^{n-m}$$

Multiplication
$$b^n \cdot b^m = b^{n+m}$$

Negative Exponent Theorem
If $x > 0$, then
$$x^{-n} = 1/x^n$$
and
$$1/x^{-n} = x^n$$

Positive integer
$$x^n = \underbrace{(x \cdot x \cdot x \ldots x)}_{n \text{ factors}}$$

Power of a power
$$(b^n)^m = b^{nm}$$

Power of a product
$$(bc)^n = b^n \, c^n$$

Power of a fraction
$$(a/b)^n = a^n/b^n$$

Zero Exponent Theorem
If b is a nonzero real number, then
$$b^0 = 1$$

Factorial Notation

Factorial notation is the use of a symbol to identify the product of a set of natural numbers. The symbol looks like this: $n!$ and is called n-factorial. The equation for n-factorial is:

$$n! = 1 \times 2 \times 3 \times \ldots \times (n-1) \times n$$

The equation when n is zero is:

$$0! = 1$$

F.O.I.L.

A popular method used to multiply two polynomials is called F.O.I.L. It stands for the order in which the multiplication is done. First, Outside, Inside, Last, and can often be calculated mentally. Here is an example of two polynomials multiplied with the F.O.I.L. method:

$$(x + 2)(x + 5)$$

F (first) $\quad = x \cdot x = x^2$
O (outside) $= 5 \cdot x = 5x$
I (inside) $\quad = 2 \cdot x = 2x$
L (last) $\quad = 5 \cdot 2 = 10$

So, this multiplies using the F.O.I.L. method as:

$$(x + 2)(x + 5) = x^2 + 5x + 2x + 10 = x^2 + 7x + 10$$

Formulas and Theorems

Area

Circle

Area = $\pi \times$ the square of the radius

$A = \pi r^2$

Parallelogram

Area = base × height

$A = bh$

Rectangle/Square

Area = length × width

$A = lw$

Rhombus

Area = (length of diagonal 1 × length of diagonal 2) ÷ 2

$A = d_1 d_2 \div 2$

Surface of a sphere

Area = $\pi \times$ square of the diameter

$A = \pi d^2$

Surface of a cylinder

Area = $2 \times \pi \times$ square of the radius plus $2 \times \pi \times$ the radius × the height

$A = 2\pi r^2 + 2\pi rh$

Trapezoid

Area = height × length of each parallel side a and b ÷ 2

$A = h(a + b) \div 2$

Triangle

Area = base × height ÷ 2

$A = bh \div 2$

Binomial Square Theorem

$(a + b)^2 = a^2 + 2ab + b^2$

and

$(a - b)^2 = a^2 - 2ab + b^2$

Circumference

$C = 2\pi r$

Diagonals in a polygon

$d = \dfrac{n(n - 3)}{2}$

(continues)

Formulas and Theorems *(continued)*

Difference of two cubes
$$a^3 - b^3 = (a - b)(a^2 + ab + b^2)$$

Difference of two squares
$$a^2 - b^2 = (a + b)(a - b)$$

Direct variation formula
$$y = kx^n$$

Distance formula between two points
$$d = \sqrt{(x_2 - x_1)^2 + (y_2 - y_1)^2}$$

Distance formula in three dimensions
$$d = \sqrt{(x_1 - x_2)^2 + (y_1 - y_2)^2 + (z_1 - z_2)^2}$$

Fibonacci sequence
For $n \geq 3$
$$t_1 = 1$$
$$t_2 = 1$$
$$t_n = t_n - 1 + t_n - 2$$

Hyperbola (inverse variation)
$k \neq 0$
$$y = k/x$$

Inverse-square variation
$k \neq 0$
$$y = k/x^2$$

Linear equations
$$ax + b = 0$$

Linear equation in two terms
$$ax + by + c = 0$$

Linear equation in three terms
$$ax + by + cz + d = 0$$

Pascal's triangle
$$\binom{n}{r} = \frac{n!}{r!(n - r)!}$$

(continues)

Formulas and Theorems *(continued)*

Perfect Square Trinomial

$a^2 + 2ab + b^2$

Perimeter

Circle

See CIRCUMFERENCE.

Equilateral Triangle

Perimeter = length of any side $\times$ 3

$p = 3s$

Rectangle

Perimeter = length $\times$ 2 plus the width $\times$ 2

$p = 2l + 2w$

Regular polygon

Perimeter = the number of n sides $\times$ length of any side

$p = ns$

Square

Perimeter = length of any side $\times$ 4

$p = 4s$

Triangle

Perimeter = total length of side a + side b + side c

$p = a + b + c$

Point slope theorem

$y - y_1 = m(x - x_1)$

Pythagorean theorem

The hypotenuse squared = the length of side a squared plus side b squared

$c^2 = a^2 + b^2$

Quadratic Equations

Quadratic Formula Theorem

If $ax^2 + bx + c = 0$, and $a \neq 0$, then

$$x = \frac{-b \pm \sqrt{b^2 - 4ac}}{2a}$$

Standard form of a quadratic equation

$ax^2 + bx + c = 0$

Standard form of a quadratic equation in two variables

$y = ax^2 + bx + c$

(continues)

Formulas and Theorems *(continued)*

Rate of gravity

$$g = 32 \ {}^{\text{ft}}/_{\text{sec}}{}^2$$

Slope

$$\frac{y_2 - y_1}{x_2 - x_1}$$

Slope-intercept

$$y = mx + b$$

Sum of two cubes

$$a^3 + b^3 = (a + b)(a^2 - ab + b^2)$$

Fractions

A fraction is any expression that is written as two quantities in which one is divided by the other. For example, $\frac{1}{2}, \frac{3}{4}, \frac{5}{8}, \frac{7}{16}, \frac{14}{x + y}$.

There are several things to keep in mind when working with fractions:

Adding fractions

If the denominators are the *same* (common denominators), add the numerators.

$$\frac{a}{b} + \frac{c}{b} = \frac{a + c}{b}$$

$$\frac{2}{4} + \frac{3}{4} + \frac{1}{4} = \frac{6}{4}$$

If the denominators are *different,* convert the fractions to higher terms to give them the same denominator, then add the fractions together.

$$\frac{2}{4} + \frac{1}{3} = \frac{6}{12} + \frac{4}{12} = \frac{10}{12} = \frac{5}{6}$$

(continues)

Fractions *(continued)*

If the denominators are different and there are only two fractions to add, another method is to multiply the numerator of each fraction by the denominator of each fraction, and multiply each of the denominators.

$$\frac{a}{b} + \frac{c}{d} = \frac{ad + bc}{bd}$$

$$\frac{2}{4} + \frac{1}{3} = \frac{2 \times 3 + 1 \times 4}{4 \times 3} = \frac{6 + 4}{12} = \frac{10}{12} = \frac{5}{6}$$

Canceling

Finding the like factors in the numerator and the denominator and deleting them to find the lowest terms. For example:

$$\frac{18}{48} = \frac{3 \times 6}{8 \times 6} \text{ cancel out each 6 from the numerator and denominator} = \frac{3}{8} \text{ so,}$$

$$\frac{18}{48} = \frac{3}{8}$$

Common denominator   The same denominator is in two or more fractions.

Complex fraction   The numerator, the denominator, or both contain a fraction. For example:

Denominator   The number on the bottom, below the division line.

Dividing fractions

Invert the second fraction, then multiply the fractions (see above for multiplication of fractions).

$$\frac{a}{b} \div \frac{c}{d} = \frac{a}{b} \times \frac{d}{c} = \frac{ad}{bc}$$

$$\frac{7}{8} \div \frac{2}{3} = \frac{7}{8} \times \frac{3}{2} = \frac{7 \times 3}{8 \times 2} = \frac{21}{16}$$

Dividing Complex Fractions

Convert the fractional bar into a division sign.

$$\frac{\frac{1}{2}}{7} = \frac{1}{2} \div 7 = \frac{1}{2} \times \frac{1}{7} = \frac{1}{14}$$

(continues)

Fractions *(continued)*

Fractional Exponents
Convert to a radical, in which the numerator becomes the power and
the denominator becomes the root.
$$x^{n/b} = (\sqrt[b]{x})^n = \sqrt[b]{x^n}$$

Highest common factor   The biggest number that can be factored out of both the numerator and the denominator.

Improper fraction   The numerator is larger than the denominator.

Like fractions   Fractions with the same denominator.

Lowest terms   All common factors in both the numerator and denominator are canceled.

Multiplying fractions
Multiply the numerators and multiply the denominators, then simplify the terms.
$$\frac{a}{b} \times \frac{c}{d} = \frac{ac}{bd}$$
$$\frac{7}{8} \times \frac{2}{3} = \frac{7 \times 2}{8 \times 3} = \frac{14}{24} = \frac{7}{12}$$

Negative signs in fractions
The negative sign can be placed in either the numerator, the denominator, or in front of the entire fraction. In any case, the fraction will have a negative value. For example:
$$\frac{-3}{4}$$
$$-\frac{2}{3}$$
$$\frac{7}{-8}$$

Numerator   The number on the top, above the division line.

Proper fraction   The numerator is smaller than the denominator.

(continues)

Fractions *(continued)*

Reciprocal

The numerator and denominator are inverted. For example:

Fraction	Reciprocal
$^1/_3$	$^3/_1$
$^2/_5$	$^5/_2$
$^3/_4$	$^4/_3$
$^6/_8$	$^8/_6$

Subtracting fractions

If the denominators are the *same,* subtract the numerators.

$$\frac{a}{b} - \frac{c}{b} = \frac{a-c}{b}$$

$$\frac{5}{12} - \frac{3}{12} = \frac{2}{12}$$

If the denominators are *different,* convert the fractions to higher terms to give them the same denominator, then subtract the fractions.

$$\frac{3}{4} - \frac{2}{3} = \frac{9}{12} - \frac{8}{12} = \frac{1}{12}$$

If the denominators are different and there are only two fractions to subtract, another method is to multiply the numerator of each fraction by the denominator of each fraction, *and* multiply each of the denominators.

$$\frac{a}{b} - \frac{c}{d} = \frac{ad - bc}{bd}$$

$$\frac{2}{4} - \frac{1}{3} = \frac{2 \times 3 - 1 \times 4}{4 \times 3} = \frac{6-4}{12} = \frac{2}{12} = \frac{1}{6}$$

Geometric Sequence and Common Ratio

A geometric sequence is a set of numbers or terms that has a pattern running through it, and that pattern has a common ratio. When one number or term in a geometric sequence is divided into the number or term that immediately follows it, the result is always the same, and that result is called the common ratio. Here are some geometric sequences, and their common ratios:

Geometric Sequence	Common Ratio
2, 4, 8, 16, 32, _____	2

$4 \div 2 = 2$
$8 \div 4 = 2$
$16 \div 8 = 2$
$32 \div 16 = 2$

3, 9, 27, 81, 243, _____	3

$9 \div 3 = 3$
$27 \div 9 = 3$
$81 \div 27 = 3$
$243 \div 81 = 3$

1, 4, 16, 64, 256, _____	4

$4 \div 1 = 4$
$16 \div 4 = 4$
$64 \div 16 = 4$
$256 \div 64 = 4$

5, 25, 125, 625, 3125, _____	5

$25 \div 5 = 5$
$125 \div 25 = 5$
$625 \div 125 = 5$
$3125 \div 625 = 5$

2, 12, 72, 432, 2592, _____	6

$12 \div 2 = 6$
$72 \div 12 = 6$
$432 \div 72 = 6$
$2592 \div 432 = 6$

Inverse Operations

Inverse operations are often used to check answers for accuracy. For example, addition can be used to check subtraction, and multiplication can be used to check division.

$$
\begin{array}{c}
25 \\
+\,39 \\
\hline
64
\end{array}
\qquad
\begin{array}{c}
\text{use inverse operations} \\
\text{to check for accuracy}
\end{array}
\qquad
\begin{array}{c}
64 \\
-\,39 \\
\hline
25
\end{array}
$$

Measurements

Arc and Angular

Value	Is the same as
Second	*"*
60 seconds	1 minute
Minute	*'*
60 minutes	1 degree
Degree	°
Right angle	90° angle
Quadrant	90° arc
Straight angle/straight line	180°
Circumference/4 right angles	360°

Capacity (and Weight), Dry

Value	Is the same as
2 pints	1 quart
1 quart dry	67.20 cubic inches
8 quarts	1 peck
4 pecks	1 bushel
1 bushel	2150.42 cubic inches

Capacity (and Weight), Liquid

Value	Is the same as
8 ounces	1 cup
16 ounces	1 pint
2 cups	1 pint

(continues)

Measurements *(continued)*

Capacity (and Weight), Liquid *(continued)*

Value	*Is the same as*
4 gills	1 pint
2 pints	1 quart
1 quart	57.75 cubic inches
4 quarts	1 gallon (231 cubic inches)
1 gallon water	8.5 pounds
7.5 gallons water	1 cubic foot
7.5 gallons water	62.5 pounds
31.5 gallons	1 barrel
1.0567 quarts	1 liter
10 liters	1 dekaliter
10 dekaliters	1 hectoliter
100 liters	1 hectoliter
1,000 liters	1 kiloliter (1,056 quarts)

Counting

Value	*Is the same as*
2 units	1 pair
12 units	1 dozen
13 units	1 baker's dozen
20 units	1 score
12 dozen	1 gross
12 gross	1 great gross

Cubic

Value	*Is the same as*
1 cubic inch	16.387 cubic centimeters
1,728 cubic inches	1 cubic foot
1 cubic foot	0.028317 cubic meter
27 cubic feet	1 cubic yard
128 cubic feet	1 cubic cord of wood (8 ft. long $\times$ 4 feet wide $\times$ 4 feet high)
1 cubic yard	0.76455 cubic meter
1.3079 cubic yards	1 cubic meter

(continues)

Measurements *(continued)*

Cubic *(continued)*

Value	*Is the same as*
0.23990 cubic mile	1 cubic kilometer
1 cubic mile	4.16818 cubic kilometers
1,000 cubic millimeters	1 cubic centimeter
1,000 cubic centimeters	1 cubic decimeter
1,000 cubic decimeters	1 cubic meter
0.061023 cubic inch	1 cubic centimeter
61.023 cubic inches	1 cubic decimeter
35.315 cubic feet	1 cubic meter

Length and Distance

Value	*Is the same as*
4 inches	1 hand (for measuring height of horses)
12 inches	1 foot
3 feet	1 yard
6 feet	1 fathom (for measuring depth of the sea)
$5\frac{1}{2}$ yards ($16\frac{1}{2}$ feet)	1 rod
40 rods ($\frac{1}{8}$ mile)	1 furlong
8 furlongs	1 mile
320 rods	1 mile
5,280 feet	1 mile
knot/nautical mile	6,080.27 feet (for measuring distance at sea)
knot/nautical mile	1.15 linear miles
10 millimeters	1 centimeter
10 centimeters	1 decimeter
10 decimeters	1 meter
10 meters	1 decameter
10 decameters	1 hectometer
10 hectometers	1 kilometer
1 inch	25.4 millimeters
1 inch	2.54 centimeters
1 foot	30.48 centimeters
1 foot	3.048 decimeters
1 foot	0.3048 meter
1 yard	0.9144 meter

(continues)

Measurements *(continued)*

Length and Distance *(continued)*

Value	Is the same as
1 mile	1609.3 meters
1 mile	1.6093 kilometers
0.03937 inch	1 millimeter
0.3937 inch	1 centimeter
3.937 inches	1 decimeter
39.37 inches	1 meter
3.2808 feet	1 meter
1.0936 yards	1 meter
1093.6 yards	1 kilometer
0.62137 mile	1 kilometer

Square

Value	Is the same as
144 square inches	1 square foot
9 square feet	1 square yard
30.25 square yards	1 square rod
160 square rods	1 square acre
640 square acres	1 square mile
1 square mile	Section
6 square miles/36 sections	1 township
1 square inch	645.16 square millimeters
1 square inch	6.4516 square centimeters
1 square foot	929.03 square centimeters
1 square foot	9.2903 square decimeters
1 square foot	0.092903 square meter
1 square yard	0.83613 square meter
1 square mile	2.5900 square kilometers
0.0015500 square inch	1 square millimeter
0.15500 square inch	1 square centimeter
15.500 square inches	1 square decimeter
0.10764 square foot	1 square decimeter
1.1960 square yards	1 square meter
0.38608 square mile	1 square kilometer

(continues)

Measurements *(continued)*

Weight (Avoirdupois and Metric)

Value	*Is the same as*
Grain	1/7000 of a pound
16 ounces	1 pound
0.035274 ounce	1 gram
1 ounce	28.350 grams
1 pound	7000 grains
1 pound	0.45359 kilogram
1 hundredweight	100 pounds (45.36 kilograms)
1 ton	2000 pounds
1 long ton	2240 pounds
2.2046 pounds	1 kilogram
2204.62 pounds	1 metric ton
10 milligrams	1 centigram
10 centigrams	1 decigram
10 decigrams	1 gram
10 grams	1 decagram
10 decagrams	1 hectogram
10 hectograms	1 kilogram
100 kilograms	1 quintal
10 quintals	1 metric ton
1000 kilograms	1 metric ton

Weight, Troy (metals)

Value	*Is the same as*
24 grains	1 pennyweight
20 pennyweights	1 ounce
12 ounces	1 pound

Weight, Carat

Value	*Is the same as*
$3^1/_2$ Troy grains	1 carat (stones and gems)
1 carat gold	1/24 gold in metal alloy
18 carats	18/24 gold in metal alloy (3/4 pure gold)
24 carats	100% pure gold

(continues)

Measurements *(continued)*

Apothecaries' Weight

Value	*Is the same as*
20 grains	1 scruple
3 scruples	1 dram
8 drams	1 ounce
12 ounces	1 pound

Apothecaries' Liquid Measure

Value	*Is the same as*
60 drops	1 fluid dram
8 fluid drams	1 fluid ounce
16 fluid ounces	1 pint
8 pints	1 gallon

Time

Value	*Is the same as*
60 seconds	1 minute
60 minutes	1 hour
24 hours	1 day
7 days	1 week
365 days	1 year
365 days, 5 hours, 48 minutes, 46 seconds	1 solar year
366 days (note: years divisible by 4, and centennial years divisible by 400, are leap years)	1 leap year
10 years	1 decade
20 years	1 score
100 years	1 century
1,000 years	1 millennium

(continues)

Measurements *(continued)*

Temperature

Fahrenheit

Water freezes	32°F
Water boils	212°F
Normal body temperature	98.6°F

Fahrenheit = (C × 1.8) + 32°

Celsius

Water freezes	0°C
Water boils	100°C
Normal body temperature	37°C

Celsius = (F − 32°) ÷ 1.8

Metric Prefixes

atto-	0.000,000,000,000,000,001 (quintillionth)
femto-	0.000,000,000,000,001 (quadrillionth)
pico-	0.000,000,000,001 (trillionth)
nano-	0.000,000,001 (billionth)
micro-	0.000,0001 (millionth)
milli-	0.001
centi-	0.01
deci-	0.1
deca-	10
hecto-	100
kilo-	1,000
myria-	10,000
mega-	1,000,000 (million)
giga-	1,000,000,000 (trillion)
tera-	1,000,000,000,000 (billion)
peta-	1,000,000,000,000,000 (quadrillion)
exa-	1,000,000,000,000,000,000 (quintillion)

Monomial

A monomial is simply any expression that consists of just one term. When monomials are put together in a string to form an expression with more than one term, the expression is then called a polynomial. *See also* POLYNOMIAL *in this appendix for more information.*

Here are some examples of monomials:

$\frac{1}{2}$

3

$4x$

$5xy$

$6y^2$

$7xy^3$

$8x^2y^3$

$\sqrt{9}$

π

Numbers

Composite numbers

Any number divisible by itself and 1, plus at least one additional positive integer.

4
6
8
9
10
12
14
15
16
18
20
etc.

(continues)

Numbers *(continued)*

Even numbers
Any number exactly divisible by 2. Any number that ends in 0, 2, 4, 6, 8.

Imaginary number
The square root of any negative real number.

Integers
The set of integers includes any whole number, and can be negative, positive, or zero.

   {. . . –5, –4, –3, –2, –1, 0, 1, 2, 3, 4, 5 . . .}

The names of integers as they are represented to the left of a decimal point are:

1.	Units
10.	Tens
100.	Hundreds
1,000	Thousands
10,000.	Ten thousands
100,000.	Hundred thousands
1,000,000.	Millions
10,000,000.	Ten millions
100,000,000.	Hundred millions
1,000,000,000.	Billions
10,000,000,000.	Ten billions
100,000,000,000.	Hundred billions
1,000,000,000,000.	Trillions (a million millions, 10^{12})
1,000,000,000,000,000.	Quadrillion (10^{15})
1,000,000,000,000,000,000.	Quintillion (10^{18})
	Sextillion (10^{21})
	Septillion (10^{24})
	Octillion (10^{27})
	Nonillion (1,000 octillions, 10^{30})

Irrational numbers
Any real number that is not a rational number. Irrational numbers cannot be written as a simple fraction.

π
$\sqrt{2}$

(continues)

Numbers *(continued)*

Natural numbers

The set of natural numbers includes any positive whole numbers beginning with the number 1.

{1, 2, 3, 4, 5, 6, 7, 8, 9, 10, 11, 12, ...}

Odd numbers

Any number not divisible exactly by 2.

Prime numbers

Any number divisible only by itself and 1.

2
3
5
7
11
13
17
19
23
29
etc.

Rational numbers

Any number, whether an integer or a fraction, which can be written as a fraction, as long as zero is not the denominator.

Real number

Any number, whether rational or irrational, which can be written with a decimal and is not an imaginary number.

Whole numbers

The set of whole numbers includes any number that is not a fraction, beginning with zero.

{0, 1, 2, 3, 4, 5, 6, 7, 8, 9, 10, 11, 12, ...}

Operations of Squares

Multiplication

$\sqrt{a} \times \sqrt{b} = \sqrt{a \times b}$

For example, $\sqrt{4} \times \sqrt{9} = \sqrt{4 \times 9} = \sqrt{36}$

$\sqrt{ab} = \sqrt{a} \times \sqrt{b}$

For example, $\sqrt{36} = \sqrt{4} \times \sqrt{9}$

Addition

$\sqrt{a} + \sqrt{a} = 2\sqrt{a}$

P.E.M.D.A.S.

The order of operations used to evaluate any expression is called P.E.M.D.A.S. An easy way to remember the order of operations is "**P**lease **E**xcuse **M**y **D**ear **A**unt **S**ally," and it stands for doing the math in this order:

Parentheses
Exponents
Multiplication
Division
Addition
Subtraction

It is important to remember to always do the multiplication and division before doing any addition and subtraction, so when both multiplication and addition are inside the parentheses, the multiplication needs to be done first. Here is an example of an expression evaluated using P.E.M.D.A.S.:

$(8 + 3 \times 5) - 4 \div 2$

Parentheses:

$(8 + 3 \times 5) - 4 \div 2$

Multiplication:

$(8 + \mathbf{3 \times 5}) - 4 \div 2$
$(8 + 15) - 4 \div 2$

Addition:

$\mathbf{(8 + 15)} - 4 \div 2$
$(23) - 4 \div 2$

Division:

$(23) - \mathbf{4 \div 2}$
$(23) - 2$

Subtraction:

$(23) - \mathbf{2} = 21$

Plane Figures

Circle—closed plane figure that is, at all points, equidistant from the center

Polygon—closed plane figure with three or more line segments

    Regular polygon—all interior angles are equal, and all line segments are equal
    Equilateral polygon—all sides are equal
    Equiangular polygon—all angles are equal

Triangle—three-sided polygon

    Equilateral—all sides are equal
    Isosceles—two sides are equal
    Isosceles right—two sides are equal and one angle is 90°
    Scalene—no sides are equal
    Right triangle—one angle is 90°
    Acute triangle—all angles are less than 90°
    Oblique-angled triangle—no right angles

Quadrilateral (tetragon)—four-sided polygon

    Trapezoid—quadrilateral with only two parallel sides
    Isosceles trapezoid—the nonparallel sides are equal in length
    Trapezium—quadrilateral with no parallel sides

Parallelogram—quadrilateral, opposite sides are parallel and equal in length

    Square—right-angled parallelogram with all four sides equal in length
    Rectangle—right-angled parallelogram
    Rhombus—oblique-angled parallelogram with four equal sides
    Diamond—quadrilateral with two obtuse angles and two acute angles

Pentagon—five-sided polygon

Hexagon—six-sided polygon

Heptagon—seven-sided polygon

Octagon—eight-sided polygon

Nonagon—nine-sided polygon

Decagon—10-sided polygon

Dodecagon—12-sided polygon

Polynomial

Any expression that consists of a string of two or more monomials is called a polynomial.

Here are some examples of polynomials:

$\frac{1}{2} + 5xy$

$8x^2y^3 + \sqrt{9} - 3$

$4x - 6y^2 + 7xy^3 - \pi$

Binomials

A polynomial that has only two terms, in other words an expression that consists of a string of just two monomials, is also called a binomial. Here are some examples of binomials:

$^1/_2 + 5xy$

$^1/_2x + 3x$

$5xy - 2xy$

$-^1/_3x + 4x^2y$

Binomial Formulas

Difference of two cubes

$\quad x^3 - y^3 = (x - y)(x^2 + xy + y^2)$

Difference of two squares

$\quad x^2 - y^2 = (x + y)(x - y)$

Sum of two cubes

$\quad x^3 + y^3 = (x + y)(x^2 - xy + y^2)$

Degree of a polynomial

The highest exponent in a polynomial is the degree of the polynomial.

$\frac{1}{2}x + 5xy^2 + \pi$ the degree is 2

$3x^2 + 4y^3$ the degree is 3

$4x^5 - 5xy^2 + 2x$ the degree is 5

Polynomials with varying degrees are often written with the highest degree first, so $3x^2 + 4y^3$ would instead be written as $4y^3 + 3x^2$.

(continues)

Polynomial *(continued)*

Like Terms

When the monomials in the string have the same variables, they are called *like terms*. Here are some examples:

$1/2x + 3x$

$5xy - 2xy$

$4x^2y^3 + x^2y^3 - 1/3\ x^2y^3$

Numbers as Polynomials

To express a number in the form of a polynomial, the number is expanded to its power of 10 and written as an expression. Here are some examples of numbers that are expanded into polynomial form:

43	$4(10^1) + 3$
747	$7(10^2) + 4(10^1) + 7$
2,556	$2(10^3) + 5(10^2) + 5(10^1) + 6$

Trinomials

A polynomial that has exactly three terms, in other words, an expression that consists of a string of three monomials, is also called a trinomial. Here are some examples of trinomials:

$8x^2y^3 + \sqrt{9} - 3$

$4xy^2 + 6y - 7x^2y$

$4x^2y^3 + x^2y^3 - {}^1/_3x^2y^3$

Unlike Terms

When the monomials in the string have different variables, they are called *unlike terms*.

Here is an example:

$-{}^1/_3x + 4x^2y$

Properties and Theorems

$1/n$ exponent theorem
  If $x \geq 0$
  $x^{1/n}$ is the nth root of x

Addition of Fractions
  If $c \neq 0$
  $\dfrac{a}{c} + \dfrac{b}{c} = \dfrac{a + b}{c}$

Addition of Like Terms
  $ac + bc = (a + b)c$

Additive Identity Property
  $0 + a = a$
  and
  $a + 0 = a$

Additive Inverse Property
  $a + -a = -a + a = 0$

Additive Property of Equality
  If $a = b$
    then $a + c = b + c$
    and $c + a = c + b$
  If $a = b$, and $c = d$
    then $a + c = b + d$
    and $c + a = d + b$

Additive Property of Inequality
  If $a < b$
    then $a + c < b + c$
    and $a - c < b - c$
  If $a > b$
    then $a + c > b + c$
    and $a - c > b - c$

Associative Property of Addition
  $(a + b) + c = a + (b + c)$

Associative Property of Multiplication
  $(a \times b) \times c = a \times (b \times c)$

(continues)

Properties and Theorems *(continued)*

Axiom of Comparison
  Only one of the following conditions can be true:
    $a < b$, or
    $a = b$, or
    $b < a$

Binomial Square Theorem
  $(a + b)^2 = a^2 + 2ab + b^2$
  and
  $(a - b)^2 = a^2 - 2ab + b^2$

Closure Property of Addition
  $a + b$ is a real number

Closure Property of Multiplication
  ab is a real number

Commutative Property of Addition
  $a + b = b + a$

Commutative Property of Multiplication
  $ab = ba$

Discriminate Theorem
  If a, b, and c are real numbers, and $a \neq 0$,
    then $ax^2 + bx + c = 0$
    has 0 real roots if $b^2 - 4ac < 0$
    has 1 real root if $b^2 - 4ac = 0$
    has 2 real roots if $b^2 - 4ac > 0$

Distributive Property of Multiplication over Addition
  $a(b + c) = ab + ac$
  and
  $(a + b)c = ac + bc$

Distributive Property of Multiplication over Subtraction
  $a(b - c) = ab - ac$
  and
  $(a - b)c = ac - bc$

Division of Real Numbers
  If $b \neq 0$
    $a \div b = a \times 1/b$

(continues)

Properties and Theorems *(continued)*

Identity Property
$$a = a$$

Identity Property of Addition
$$a + 0 = 0 + a = a$$

Identity Property of Multiplication
$$a \times 1 = 1 \times a = a$$

Inverse Property of Addition
$$a + {-a} = {-a} + a = 0$$

Inverse Property of Multiplication
If $a \neq 0$
$$a \times 1/a = 1/a \times a = 1$$

Multiplicative Identity Property
$$a \times 1 = a$$
and
$$1 \times a = a$$

Multiplicative Property of Equality
If $a = b$
then $ac = bc$
and $ca = cb$
If $a = b$ and $c = d$
then $ac = dc$
and $ac = bc$

Multiplicative Property of Inequality
If $a < b$, and $c > 0$
then $ac < bc$
and $a/c < b/c$
If $a < b$ and $c < 0$
then $ac > bc$
and $a/c > b/c$

Multiplicative Property of Zero
$$a \times 0 = 0$$
and
$$0 \times a = 0$$

(continues)

Properties and Theorems *(continued)*

Multiplicative Property of -1

$a \times -1 = -a$

and

$-1 \times a = -a$

nth Root of nth Power Theorem

For all real numbers x, and all integers $n \geq 2$,

If n is odd, then

$\sqrt[n]{x^n} = x$

and

if n is even, then

$\sqrt[n]{x^n} = |x|$

Negative Exponent Theorem

If $\times > 0$, then

$x^{-n} = 1/x^n$

and

$1/\mathrm{x}^{-n} = x^n$

Number of Subsets of a Set

2^n

n = number of elements in a set

Opposite of an Opposite Property (Op-Op Property)

$-(-a) = a$

Opposite of a Sum Property

$-(a + b) = -a + -b$

Perfect Square Trinomial

$x^2 + 2hx + h^2$

Point Slope Theorem

$y - y_1 = m(x - x_1)$

Power of a Fraction

$(a/b)^n = a^n/b^n$

Power of a Power Property

$(a^m)^n = a^{mn}$

Power of a Product Property

$(ab)^m = a^m b^m$

(continues)

Properties and Theorems *(continued)*

Power of a Quotient Property
  If $b \neq 0$
  $(a/b)^m = a^m/b^m$

Powers Property of Equality
  If $a = b$
  then $a^n = b^n$.

Product of Powers Property
  $a^m \times a^n = a^{m+n}$

Property of Exponentials
  If $a^x = a^y$, then $x = y$
  and
  If $a^x = b^x$, then $a = b$

Pythagorean theorem
  $a^2 + b^2 = c^2$
  and
  $c^2 = a^2 + b^2$

Quadratic Formula
  $ax^2 + bx + c = 0$

Quotient of Power Property
  If $a \neq 0$
  $a^m/a^n = a^{m-n}$

Rational Exponent Theorem
  If $x > 0$
  $x^{m/n} = (x^{1/n})^m$
  and
  $x^{m/n} = (x^m)^{1/n}$

Reflexive Property of Equality
  $a = a$

Root of a Power Theorem
  If $x > 0$, and $n \geq 2$, and m and n are integers
  $\sqrt[n]{x^m} = (\sqrt[n]{x})^m = x^{m/n}$

Root of a Product Theorem
  If $n > 1$, and x and y are positive real numbers
  $\sqrt[n]{xy} = \sqrt[n]{x} \cdot \sqrt[n]{y}$

(continues)

Properties and Theorems *(continued)*

Substitution Property of Equality
　If $a = b$, then
　a may be substituted for b in any expression
　and
　b may be substituted for a in any expression

Subtraction of Real Numbers
　$a - b = a + -b$

Symmetric Property of Equality
　If $a = b$, then $b = a$

Transitive Property of Equality
　If $a = b$, and $b = c$, then $a = c$

Transitive Property of Inequality
　If $a < b$, and $b < c$, then $a < c$
　　and
　If $a > b$, and $b > c$, then $a > c$

Trichotomy Property of Inequality
　Either $a < b$, $a = b$, or $a > b$

Zero Exponent Theorem
　If b is a nonzero real number, then
　$b^0 = 1$

Zero Product Theorem
　If $a = 0$ or $b = 0$, then $ab = 0$

Roman Numerals

Arabic	Roman
1	I
5	V
10	X
50	L
100	C
500	D
1,000	M
5,000	$\overline{V}$
10,000	$\overline{X}$
1,000,000	$\overline{M}$

There are seven Roman numerals that can represent any whole Arabic number by properly placing the numerals next to each other. There are a few simple rules that help convert the value of Roman numerals to Arabic:

Sum

If a numeral is followed by another numeral that is of *lesser* value, they are read as a sum. For example, in VI, the value of V (five) is followed by the value of I (one), and their overall value is the sum of five and one, or six.

VI	6
XVI	16
XXVI	26
LXVI	66

Difference

If the numeral is followed by another numeral that is of *greater* value, then they are read as a difference. For example, in IV, the value of I (one) is followed by the value of V (five), and their overall value is the difference, or four.

IV	4
IX	9
XL	40
XC	90

(continues)

Roman Numerals *(continued)*

Repeating

If a numeral repeats itself, the value is repeated. II is 2, or two times one. XX is 20, or two times 10. CC is 200, or two times 100. The Roman numerals L, V, and D are not written in a repeating sequence because LL is the same value as C (50 times 2 is the same as 100), VV is the same as X (5 times 2 is the same as 10), and DD is the same as M (500 times 2 is the same as 1,000).

II	2
III	3
XX	20
XXX	30
MM	2,000
MMM	3,000

Bar

If the Roman numeral has a bar above it, the value increases by 1,000. $\overline{V}$ is 5,000, $\overline{IV}$ is 4,000, etc.

Signed Numbers

Absolute Value

The absolute value of a signed number is the number without the sign. For example:

$|-5| = 5$

$|+27| = 27$

Adding Signed Numbers

Adding two positives equals a positive.

Adding two negatives equals a negative.

To add a positive and negative:

    a. Make both numbers absolute values

    b. Subtract the smaller number from the larger number

    c. Put the sign of the larger original number in front of the difference

For example:

    $5 + -12$

    a. $|5| + |-12|$

    b. $12 - 5 = |7|$

    c. -7

(continues)

Signed Numbers *(continued)*

Dividing Signed Numbers

Dividing two like signs, the quotient is positive.

Positive ÷ Positive = Positive
Negative ÷ Negative = Positive

Dividing two unlike signs, the quotient is negative.

Positive ÷ Negative = Negative

Multiplying Signed Numbers

Multiplying two like signs, the product is positive.

Positive × Positive = Positive
Negative × Negative = Positive

Multiplying two unlike signs, the product is negative.

Positive × Negative = Negative

Subtracting Signed Numbers

Subtracting a signed number equals adding its opposite. For example:

$4 - (-6) = 4 + 6 = 10$
$-3 - (-2) = -3 + 2 = -1$
$-9 - 7 = -9 + (-7) = -16$

Solid Figures

Solid figures are very similar to their counterparts, plane figures. All solids are closed three-dimensional shapes, many using polygons to form each face of the solid.

Sphere—all points are equidistant from the center

Paraboloid—a solid parabola

Pyramid—triangular faces on polygon base form a common vertex on top

Tetrahedron—four-sided, all faces are equilateral triangles

Hexahedron—six-sided, all faces are equal squares; a cube

Octahedron—eight-sided, all faces are equilateral triangles

Dodecahedron—12-sided, all faces are pentagons

Icosahedron—20-sided, all faces are equilateral triangles

Symbols

Addition (plus)	$+$
Subtraction (minus)	$-$
Plus or minus	$\pm$
Minus or plus	$\mp$
Multiplication (times)	$\times$, $\bullet$, $*$
	$(a)(b)$, $a(b)$, $(a)b$, ab
Division (divided by)	$\div$, $/$
Ratio	$:$
Equals	$=$
or, is equal to	
Equal and parallel	$\#$
Is approximately equal to	$\cong$, $\approx$
or, is nearly equal to	
Does not equal	$\neq$
or, is not equal to	
Identical with,	$\equiv$
congruent, or equivalent to	
Not equivalent to	$\not\equiv$, $\neq$
Less than	$<$
Less than or equal to	$\leq$, $\leqq$, $\lesssim$
Equal to or less than	$\gtreqless$
Is not less than	$\not<$
Greater than	$>$
Greater than or equal to	$\geq$, $\geqq$, $\gtrsim$
Equal to or greater than	$\lesseqgtr$
Is not greater than	$\not>$
Greater than or less than	$\gtrless$
Is approximately	$\simeq$
For every/all	$\forall$
There exists	$\exists$
There does not exist	$\nexists$
Therefore	$\therefore$
Because	$\because$
Function	f
Inverse of a function	f^1
Base of natural log	e
Area	A
Diameter	d

(continues)

Symbols *(continued)*

Circumference	C
Radius	r
Plane	P
Squared	^{2}
Cubed	^{3}
To the *n*th power	n
Square root, radical	$\sqrt{\ }, \sqrt{}$
Negative square root	$-\sqrt{\ }, -\sqrt{}$
$\sqrt{-1}$	i
Cube root	$\sqrt[3]{\ }, \sqrt[3]{}$
Fourth root	$\sqrt[4]{\ }, \sqrt[4]{}$
*n*th root	$\sqrt[n]{\ }, \sqrt[n]{}$
Least common denominator	lcd
Least common multiple	lcm

Grouping symbols

braces	{ }
brackets	[ ]
parentheses	( )

Sets

Such that	|
Set of elements	{ }
The set of all x such that…	$\{x\mid \ldots\}$
Is an element of a set	$\in$
or, belongs to a set	
Is not an element of a set	$\notin$
or, does not belong to a set	
Subset	$\subseteq$
Proper subset	$\subset$
Empty Set	$\varnothing$
or, null set	
Universal set	U
Union	$\cup$
Intersection	$\cap$

(continues)

Symbols *(continued)*

Graphs

Origin	*O*
Ordered pair	*(x, y)*
Ordered triple	*(x, y, z)*

Boolean algebra

Conjunction (and)	$\wedge$
e.g., *p* and *q*	$p \wedge q$
Disjunction (or)	$\vee$
p or *q*	$p \vee q$
Conditional (if, then)	$\rightarrow$
if *p*, then *q*	$p \rightarrow q$
Equivalence (if and only if)	$\leftrightarrow$
p if and only if *q*	$p \leftrightarrow q$
Negation (not)	$'$
not *p*	p'

Degrees	$^\circ$
Celsius	$^\circ$C
Fahrenheit	$^\circ$F
Percent	$\%$
Infinity	∞
Pi	π
Variables	*x, y, z, a, b, c*
Sum (sigma)	Σ
Angle	$<, >, \measuredangle$
Right angle	$\llcorner$
Acute angle	$\wedge, \vee$
Obtuse angle	$\frown, \smile$
Triangle	$\triangle$
Circle	$\circ$
Arc	$\cap, \cup$
Perpendicular	$\perp, \vdash$
Parallel	$\parallel$
Factorial	$!$
n-factorial	$n!$
Absolute value	$|\,|$
Integer	*n*
Real number	*R*

(continues)

Symbols *(continued)*

Measurement Abbreviations

Area	*A*
Volume	*V*
length	*l*
width	*w*
height	*h*

length

inches	in., "
feet	ft, '
yard(s)	yd
rod(s)	rd
mile(s)	mi
kilometer (1,000 meters)	km
hectometer (100 meters)	hm
decameter (10 meters)	dam
decimeter	dm
centimeter	cm
millimeter	mm

square

square inch(es)	sq. in., "
square foot/feet	sq. ft., '
square yard	sq. yd.
square rod	sq. rd.
square acre	A
square mile	sq. mi.

cubic

cubic inch(es)	cu. in.
cubic foot/feet	cu. ft.
cubic yard	cu. yd.
cubic cord	cd.

weight

ounce(s)	oz., ozs.
pound(s)	lb., lbs.
hundredweight	cwt

(continues)

Symbols *(continued)*

weight (continued)

ton(s)	ton
milligram	mg
centigram	cg
decigram	dg
gram	g
decagram (dekagram)	Dg, dag, dkg
hectogram	Hg, hg
kilogram	Kg, kg
myriagram	Mg.
Metric quintal	Q., q, ql
Metric ton	M.T.
distance	*d*
time	*t*
seconds	*sec,* "
minute	min, '
miles per hour	*mph*

Capacity

cups	c
pints	pt
quarts	qt
gallons	gal.
milliliters	mL, ml
centiliter	cl
deciliters	dl
liter	l, lit.
decaliter (dekaliter)	Dl, dal, dkl
hectoliter	Hl, hl
kiloliter	Kl, kl

Sound

bel	B
decibel	dB

Triangles

A triangle is a polygon that has three sides and three angles. The sum of the angles of a triangle always equals 180°.

Right triangle—one 90° angle
Isosceles triangle—two sides that are equal in length
Isosceles right triangle—two sides equal in length, plus one 90° angle
Equilateral triangle—all three sides are equal in length
Scalene triangle—no two sides are equal in length
Oblique-angled triangle—no right angles
Acute triangle—all angles measure less than 90°

The *angles* of a triangle are signified with capital letters *A, B,* and *C*.

The *sides* of a triangle are signified with lowercase letters *a, b,* and *c*.

The *side* of a triangle that is opposite the *angle* is signified with the same letter. For example, the side opposite angle *A* is side *a*. The side opposite angle *B* is side *b*. The side opposite angle *C* is side *c*.

The *hypotenuse* of a right triangle is the side opposite the right angle. This side is also always the longest side of a right triangle. The remaining sides are called *legs*.

The *Pythagorean Theorem* is used to determine the length of a side of a right triangle. It states that the square of the length of the hypotenuse equals the sum of the squared lengths of the other two sides. This is written as $a^2 + b^2 = c^2$.

30° – 60° – 90° triangle: A right triangle in which one of the other angles equals 30°, and the remaining angle equals 60°.

The lengths of the sides of a 30° – 60° – 90° triangle are:

Half the length of the hypotenuse on the side opposite the 30° angle.
This is written as $a = c/2$.
Half the length of the hypotenuse times $\sqrt{3}$ on the side opposite the 60° angle.
This is written as $b = c/2 \times \sqrt{3}$.

An *isosceles right triangle:*

Has angles measuring 45° – 45° – 90°.
$A = B = 45°$.
$C = 90°$.

Has two sides that are equal in length.
This is written as $a = b$.

The lengths of the two equal sides are half the hypotenuse times $\sqrt{2}$.
This is written as $b = c/2 \times \sqrt{2}$, where $a = b$.

Variables

Variables are used in algebra to express an unknown number or a quantity, and are usually represented by letters. Some of the most common variables used in algebra are *n, x,* and *y.* RENÉ DESCARTES invented the idea of using the letters at the beginning of the alphabet for known quantities, and the letters at the end of the alphabet for unknown quantities. Here are some examples of terms with variables (in each case, the variable is the letter):

x

$5y$

a^2

$6b^3$

9^n

Volume

Volume is the space occupied by a three-dimensional object.
Here is how to calculate volume *(V):*

Cylinder

Volume = 2 × π × the radius (cubed)
$V = 2\,\pi\,r^3$

Prism

Volume = area of the Base × height
$V = Bh$

Pyramid

Volume = area of the Base × height ÷ 3
$V = Bh \div 3$

Rectangular Solid

Volume = length × width × height
$V = lwh$

Sphere

Volume = diameter cubed × $^1/_6$ of π
$V = d^3 \times .5236$

Recommended Reading and Useful Websites

This list of print and interactive media is just the tip of the iceberg of the hundreds of resources used in writing this book, and is meant to provide you with a glimpse of some of the available information on the art and science of mathematics. The books vary in level from empowering and easy to understand, like *Forgotten Algebra,* to more difficult traditional textbooks, to fun math experiences as in *The Number Devil: A Mathematical Adventure.* Please use the great resource of your reference librarian to find these and more books to help in your discovery. I have listed an award-winning CD-ROM, *Grade Builder, Algebra 1,* published by the Learning Company, which received praise from the press, and more importantly from teens who repeatedly reported that they could not wait to get home from school to "play algebra." And finally, the many websites listed are some of the best on in-depth research and fun things to do with math. As you venture forth into your quest for knowledge, let this be your resource for getting started. Happy exploring!

Print

Allen, Edwin Brown, Dis Maly, S. Herbert Starkey, Jr. *Vital Mathematics.* New York: Macmillan, 1944.

Ball, W. W. Rouse. *A Short Account of the History of Mathematics.* New York: Dover Publications, 1960.

Bleau, Barbara Lee. *Forgotten Algebra: A Self-Teaching Refresher Course.* Hauppauge, N.Y.: Barron's Educational Series, 1994.

Dolciani, Mary P., William Wooton, and Edwin F. Beckenback. *Algebra 1.* Boston: Houghton Mifflin Company, 1980.

Enzensberger, Hans Magnus. *The Number Devil: A Mathematical Adventure.* Trans. Michael Henry Heim. New York: Metropolitan Books, 1998.

Fuller, Gordon. *College Algebra.* 4th ed. New York: D. Van Nostrand Company, 1977.

Goldstein, Thomas. *Dawn of Modern Science: From the Arabs to Leonardo Da Vinci.* Boston: Houghton Mifflin Company, 1980.

Hogben, Lancelot. *Mathematics for the Million: How to Master the Magic of Numbers.* New York: W. W. Norton & Company, 1993.

Knopp, Paul J. *Linear Algebra: An Introduction.* New York: John Wiley & Sons, 1974.

McClymonds, J. W., and D. R. Jones. *Advanced Arithmetic.* Sacramento: California State Series, 1910.

Reid, Constance. *A Long Way from Euclid.* New York: Thomas Y. Crowell Company, 1963.

Seife, Charles. *Zero: The Biography of a Dangerous Idea.* New York: Viking Penguin, 2000.

Senk, Sharon L., Denisse R. Thompson, Steven S. Viktora, et al. *Advanced Algebra: The University of Chicago School Mathematics Project.* Glenview, Ill.: Scott Foresman and Company, 1990.

Shulte, Albert P., and Robert E. Peterson. *Preparing to Use Algebra, Third Edition, Teacher's Edition.* River Forest, Ill.: Laidlaw, 1984.

Taton, René, ed. *History of Science: Ancient and Medieval Science.* Trans. A. J. Pomerans. New York: Basic Books, 1963.

———. *History of Science: Science in the Nineteenth Century.* New York: Basic Books, 1965.

———. *History of Science: Science in the Twentieth Century.* New York: Basic Books, 1964.

Whitesitt, J. Eldon. *Boolean Algebra and Its Applications.* Mineola, N.Y.: Dover Publications, 1995.

Yount, Lisa. *A to Z of Women in Science and Math.* New York: Facts On File, 1999.

Interactive Media and Websites

5 Numbers. Program 4—The Imaginary Number. BBC. Available on-line. URL: http://www.bbc.co.uk/radio4/science/5numbers4.shtml. Accessed November 2, 2002.

Agnes Scott College. Biographies of Women Mathematicians. Available on-line. URL: http://www.agnesscott.edu/lriddle/women/women.htm. Accessed November 2, 2002.

"All About Astronomy." PBS. Available on-line. URL: http://www.pbs.org/standarddeviantstv/transcript_astronomy.html#ptolemy. Accessed November 2, 2002.

Cartage: Your Compass in Learning. Available on-line. URL: http://www.cartage.org.lb. Accessed November 2, 2002.

Center for Relativity, University of Texas. "Genius and Biographers: The Fictionalization of Evariste Galois." Available on-line. URL: http://godel.ph.utexas.edu/~tonyr/galois.html. Accessed November 2, 2002.

Central University of Venezuela Mathematics School. Famous Mathematicians. Available on-line. URL: http://euler.ciens.ucv.ve/English/mathematics. Accessed November 2, 2002.

Clark University. "Euclid's Elements." Available on-line. URL: http://aleph0.clarku.edu/~djoyce/java/elements/elements.html. Accessed November 2, 2002.

Cosmiverse: Your Universe Online. Available on-line. URL: http://www.cosmiverse.com. Accessed November 2, 2002.

Drexel University. *Archimedes.* Available on-line. URL: http://www.mcs.drexel.edu/~crorres/Archimedes/contents.html. Accessed November 2, 2002.

The Fundamental Theorem of Algebra. Available on-line. URL: http://www.und.nodak.edu/dept/math/history/fundalg.htm. Accessed November 2, 2002.

Geometry: The Online Learning Center. Available on-line. URL: http://www.geometry.net. Accessed November 2, 2002.

Grade Builder Algebra 1. Available on CD-ROM. The Learning Company, 1997.

The History Behind the Thermometer. Available on-line. URL: http://inventors.about.com/library/inventors/blthermometer.htm. Accessed November 2, 2002.

HyperHistory Online. Available on-line. URL: http://www.hyperhistory.com/online_n2/History_n2/a.html. Accessed November 2, 2002.

Indiana University. Temperature Calculator. Available on-line. URL: http://www.indiana.edu/~animal/fun/conversions/temperature.html. Accessed November 2, 2002.

Lee, J. A. N. "Charles Babbage." Available on-line. URL: http://ei.cs.vt.edu/~history/Babbage.html. Accessed November 2, 2002.

Knott, Ron. Fibonacci Numbers and the Golden Section. Surrey University. http://www.mcs.surrey.ac.uk/Personal/R.Knott/Fibonacci. Accessed November 2, 2002.

Linda Hall Library: Science, Engineering, and Technology. Available on-line. URL: http://www.lindahall.org. Accessed November 2, 2002.

Lucas, Tim. 235 AD. Available on-line. URL: http://faculty.oxy.edu/jquinn/home/Math490/Timeline/235AD.html. Accessed November 2, 2002.

MacTutor: History of Mathematics. University of St. Andrews, Scotland: School of Mathematics and Statistics. Available on-line. URL: http://www-history.mcs.st-andrews.ac.uk/history/index.html. Accessed November 2, 2002.

NASA. Athena Earth and Space Science for Kids K-12. Available on-line. URL: http://vathena.arc.nasa.gov. Accessed November 2, 2002.

National Academy of Sciences. Available on-line. URL: http://www4.nas.edu/nas/nashome.nsf?OpenDatabase. Accessed November 2, 2002.

Neo-Tech: The Philosophical Zero. "The Geocentric Hierarchy." Available on-line. URL: http://www.neo-tech.com/zero/part3.html. Accessed November 2, 2002.

Parshall, Karen Hunger. "The Art of Algebra from Al-Khwarizmi to Viète: A Study in the Natural Selection of Ideas." Available on-line. URL:

http://www.lib.virginia.edu/science/parshall/algebraII.htm#CONT. Accessed November 2, 2002.

PBS: Nova. "Math's Hidden Woman" Available on-line. URL: http://www.pbs.org/wgbh/nova/proof/germain.html. Accessed November 2, 2002.

Platt, Daniel. A History of Fermat's Last Theorem. University of Bath. Available on-line. URL: http://www.bath.ac.uk/~ma0dmp/FLTfront.html. Accessed November 2, 2002.

Rice University. The Galileo Project. Available on-line. URL: http://es.rice.edu/ES/humsoc/Galileo. Accessed November 2, 2002.

Sacklunch. Simply Biographies. Available on-line. URL: http://www.sacklunch.net/biography. Accessed November 2, 2002.

Schultz, Phil. History of Mathematics 3M3. Available on-line. URL: http://www.maths.uwa.edu.au/~schultz/3M3/2000_Course_Notes.html. Accessed November 2, 2002.

Seton Hall University. Interactive Real Analysis. Available on-line. URL: http://www.shu.edu/html/teaching/math/reals/gloss/index.html. Accessed November 2, 2002.

Simon Fraser University. British Columbia, Canada. History of Mathematics. Available on-line. URL: http://www.math.sfu.ca/histmath. Accessed November 2, 2002.

The Sixteenth-Century Mathematics of Italy: Commercial Mathematics. Available on-line. URL: http://library.thinkquest.org/22584/emh1400.htm?tqskip1=1&tqtime=0708. Accessed November 2, 2002.

The Story of Thermometer. Available on-line. URL: http://inventors.about.com/gi/dynamic/offsite.htm?site=http://www.ulearntoday.com/magazine/physics%5Farticle1.jsp%3FFILE=thermometer. Accessed November 2, 2002.

Tufts University. The Perseus Digital Library. Available on-line. URL: http://www.perseus.tufts.edu. Accessed November 2, 2002.

University College London. "De Sacro Bosco's 'De Sphera' and Other Tracts." Available on-line. URL: http://www.aim25.ac.uk/cgi-bin/search2?coll_id=1612&inst_id=13. Accessed November 2, 2002.

University of Buffalo Library. "Aristotle and Cosmology I." Available on-line. URL: http://ublib.buffalo.edu/libraries/units/sel/exhibits/stamps/cosmo1.htm. Accessed November 2, 2002.

University of California. Contributions of 20th Century Women to Physics. Available on-line. URL: http://www.physics.ucla.edu/~cwp. Accessed November 2, 2002.

University of Dublin, Trinity College School of Mathematics. Available on-line. URL: http://www.maths.tcd.ie/pub/Resources. Accessed November 2, 2002.

Uppsala Astronomical Observatory. Sweden. Available on-line. URL: http://www.astro.uu.se. Accessed November 2, 2002.

Van Helden, Albert. "Ptolemaic System." Available on-line. URL: http://es.rice.edu/ES/humsoc/Galileo/Things/ptolemaic_system.html. Accessed November 2, 2002.

Weisstein, Eric. *World of Mathematics.* Available on-line. URL: http://mathworld.wolfram.com.

———. *World of Science.* Available on-line. URL: http://scienceworld. wolfram.com. Accessed November 2, 2002.

Wilson, Fred L. "History of Science: Galileo and the Rise of Mechanism." Available on-line. URL: http://www.rit.edu/~flwstv/galileo.html. Accessed November 2, 2002.

Who What When: Interactive Historical Timelines. Available on-line. URL: http://www.sbrowning.com/whowhatwhen/index.php3?q=12&pid=604. Accessed November 2, 2002.

xrefer: the Web's Reference Engine. "Number Theory." Available on-line. URL: http://www.xrefer.com/entry/511441. Accessed November 2, 2002.